PHOTOACTIVATION
OF ISOMERS OF STABLE NUCLEI AND RECENT APPLICATIONS

AF847287

INSTITUTE OF ISOTOPE AND SURFACE
CHEMISTRY, CHEMICAL RESEARCH CENTRE HUNGARIAN
ACADEMY OF SCIENCES (IISC CRC HAS)

ÁRPÁD VERES

PHOTOACTIVATION OF ISOMERS OF STABLE NUCLEI AND RECENT APPLICATIONS

AKADÉMIAI KIADÓ, BUDAPEST

This book has been published with the financial assistance
of the Hungarian Academy of Sciences

Published by Akadémiai Kiadó
H-1519 Budapest, P.O. Box 245
www.akkrt.hu

ISBN 963 05 7865 4

© Árpád Veres, 2002

All rights reserved. No part of this book may be reproduced
by any means, or transmitted or translated
into machine language
without the written permission of the publisher.

Printed in Hungary

Preface

In the different fields of science and development it very often happens that specialists in a particular scientific field give summaries of their most important results. These summaries are very important for every aspect of further development.

For example, in the 18th century electricity was the main field of research and two extremely important discoveries (Faraday: magnetic induction; Maxwell: axiomatic system of electrodynamics) were considerable contributions with respect to furthering the development of electrodynamics. In the middle of the 19th century physicists were fascinated by the interaction between light and electricity. Later on, chemistry played an important role in examining the connection between the structure of matter and electricity. Moreover, those chemists who could use such methods to give numerical statements with reference to the character of molecules could make certain calculations concerning the mass and number of the molecules. At the beginning of the 20th century, in one of his lectures, Lord Kelvin suggested that there were two factors in the field of physics which obscured the beauty and clarity of the theory of dynamics. The first one centred upon the way in which physical elements could travel through ether; the second problem concerned the Maxwell–Boltzmann doctrine, which dealt with the scattering of energy.

In 1905 Einstein formulated the theory of relativity. This has been referred to as the "completion of classical physics". Einstein's prediction, that induced transitions must occur in quantum systems, was based on a purely thermodynamic argument, representing a general application (1917). Einstein dealt with the origins of gravity (involving problems which to this day are unresolved). New techniques, sometimes based upon entirely new principles – and each appropriate to the next spectral range – have, nevertheless, contributed to a steady progress towards greater and greater understanding. The quantum theory, quantum optics, radioactivity, nuclear reactions and models, quantum colour dynamics, heavy-ion physics, nuclear astrophysics, and laser techniques are all products of this century. In the last fifty years the "Big Sciences" and "High Technology" have been dominant, but particular aspects have only been significant in experiments if they have immediately proved to be useful in resolving certain difficult questions.

It has to be admitted that the photoexcitation of isomers represents only a small area in the field of nuclear physics; however, on a number of occasions it has been the only important method which has been able to solve some interesting tasks. In the 1980s for example, the isomer activation cross-section of $^{176}Lu^{m}$ became an important value from the point of view of creating a cosmic clock or thermometer. Two strongly different values were measured by Yoshihara [60Yo], (1000–2400 nb) and by Veres [70Ve] (23–47 nb). Therefore Norman and Kellogg [85No1] remeasured it and determined a value of 38±10 nanobarns for the cross-section of $^{176}Lu^{m}$.

Developments which have taken place – over the last fifty years or more – in the study of the gamma-excitation of nuclear isomers certainly provide a useful survey; this should be capable of providing an up-to-date account of our present knowledge. The latter is especially important for those entering the field who are unable to gain easy access to the flood of literature which has appeared in recent years.

Modern computer technology has revolutionised theoretical calculations in many fields of science and technology. Physical information of great complexity – previously ignored or crudely treated – can now be introduced in detail into photoexcitation, stellar nucleosynthesis, and gamma-laser computations. These can be treated with great precision and thoroughness. The primary motivation in the preparation of this work has been a desire to extend (where possible) the calculated and experimental cross-section rates for quite high temperatures and high densities. These extensions are obviously necessary for applications to the advanced stages of stellar evolution and to gamma-laser techniques. The work presented here will review the available experimental data on cross-sections for the nuclear interactions of the (γ, γ') reaction of stable nuclei which have isomeric states longer than 0.3 sec.

This study only intends to present the development of the photoexcitation of the isomeric states of stable nuclei and to show how this might be applied in the fields of astrophysics and laser-techniques, respectively.

The first half of the book examines the irradiation of the target → excitation mechanism → measurements of isomeric activity and the related evaluation processes. After a short history of earlier practical applications, there is a survey of the target nuclei, irradiation facilities, and experimental laboratories. There is also a review of measurement techniques, nuclear models, and calculations concerning photoexcitation and its application in astrophysics and gamma-laser (graser) techniques.

In order to make extensions to the astrophysical section, besides the models of nucleosynthesis, mention is made of abundance calculations, the nucleosynthetic effect of γ-fluxes on stellar surfaces and solar systems. This part of the book also looks at the very important effect of high temperature and the thermalization of long-lived isomers. At high temperatures the nuclear products of a given reaction act to reverse the original reaction. This is particularly true near the end of the burning of nuclear fuels. Photodisintegration rates become comparable to those for radioactive capture.

The difficulty of graser development, together with the perceived rewards of success, make the above a major scientific and technological challenge. Between 1965 and 1972, there are no reports of new attacks on the graser problem in the available professional literature. In 1981, Baldwin, Solem and Gol'danskii published an excellent review in the *Review of Modern Physics* using mostly the Mössbauer effect in nuclear transitions for generating coherent electromagnetic radiation in the 6–12 keV energy range. The next stage was the 1st International Gamma-Ray Laser Workshop (GARALAS '95) in Predeal (Romania). Two serious problems emerged from the many difficulties associated with the realisation of graser. These involve the production of very high energy radiation densities and also the pumping difficulties of the graser. In rare cases, selection rules sufficiently inhibit the coupling of the particle motion to the electromagnetic field for the energies to be stored for long times in those special nuclei. Such long-lived isomeric states are natural sources of high energy densities. The other great challenge concerns the pumping (or triggering) of sufficient nuclei in such a state that coupling can take place when desired.

The second half of this study attempts to provide integral cross-section values and other more characteristic data of the photoexcitation processes. (However, no attempt is made to achieve completeness in those cases in which the data are identical within the error bars.) It has been essential to be able to construct partial decay schemes from which the activation levels can be identified, given that this is very useful in the application of astrophysical, γ-laser, and other techniques.

The complexity of the field, of course, creates problems with respect to the level and depth of coverage, the duplication of material, and the use of notation. Different sections of this book frequently use different forms of notation in order to facilitate the reader's understanding of primary sources and also the notation that is most commonly used in the relevant literature.

The author is greatly indebted to his colleague Dr. László Lakosi for many helpful discussions, suggestions and criticisms during the preparation of the manuscript.

Budapest, 2001

The Author

Contents

1. Introduction

Photoexcitation of the isomers of nuclei is a method that allows one to get information about the interaction of photons with nuclei and about the structure of nuclei below the particle threshold. In some nuclei "pygmy" resonances have been found in the total photon scattering cross-section of these energies, and these resonances can be considered as a possible manifestation of some form of intermediate structure.

Photonuclear excitation of a ground-state nucleus in its metastable state has been studied extensively over a long period. This process is considered as a two-step one: first, the nucleus is excited to a so-called activation level, or activation state (Al, or As) by resonance absorption; in the next step the excited nucleus moves, by direct or cascade transition, down to the isomeric state (level) via various de-excitation branches. This process is also known in the literature as the (γ, γ') reaction.

The process of photoexcitation can take place with radioactive isotopes and also with the stable nuclei of elements occurring in nature. The latter case will be dealt with in this study.

1.1. Short history

The occurrence of the isomeric state in the artificial radioactive nuclei was first observed by Hahn [21Ha]. He investigated the β-decay of ^{234}Th and found two decay products with different half-lives (1.17 minutes and 6.7 hours, respectively). In conformity with the β-decay, the daughter product must be ^{234}Pa. This moves towards the ground state and a small part of the decay transits to an excitation state of ^{234}Pa; from here it goes to the ground state (see *Fig. 1*). This pair of nuclei was the first and, for a long time, the only known example of nuclear isomerism. After the discovery of artificial radioactivity in 1934, Kurchatov et al. [35Ku] detected two different radioactivities – with 4.4-hour and 16-minute life-times, respectively – in the ^{79}Br(n, γ)^{80}Br reaction. In the period following this new isomers were found at an increasing rate.

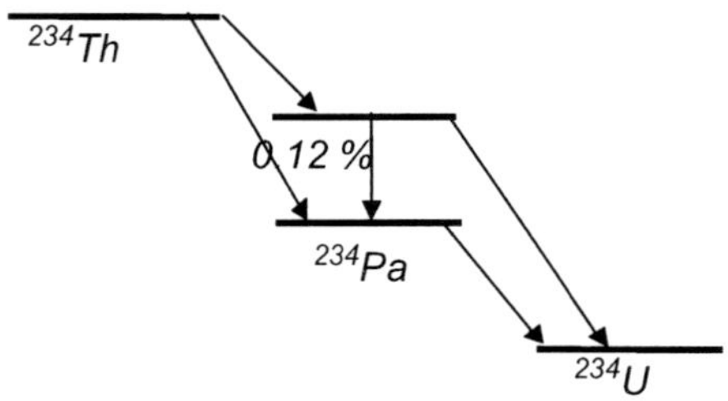

Fig. 1. Decay scheme of ^{234}Pa

According to von Weizsäcker [36We], nuclear isomerism may occur whenever the angular momentum of a low-lying state of the nucleus differs from the angular momentum of any lower state by several units of $h/2\pi$.

The first conclusive evidence of nuclear isomerism in a stable nucleus was reported by Goldhaber et al. [38Go]. They found that the 4.1-hour β^- activity of indium could be produced by fast neutrons but not by slow neutrons; in fact, in this case, the process seemed to have a threshold.

The X-ray excitation of the isomeric state of a stable ^{115}In nucleus was first observed by Pontecorvo and Lazard [39Po], and by the Notre Dame group [39Co]. They reported a threshold of 1.2 ± 0.1 MeV. In 1943, Korsunsky et al. [43Ko] found the threshold at 1.07 ± 0.05 MeV; this was in good agreement with the value of the Notre Dame group. The cross-section was estimated to be 10^{-33} cm^2. These results agree with the Bethe–Heitler theory, but this theory does not hold near the threshold. Waldman and Wiedenbeck [43Wa] proved the possibility of excitation at higher energy and found evidence for activation levels at 1.12 MeV, 1.55 MeV, 2.13 MeV, and 2.64 MeV.

Goldhaber et al. [39Go] irradiated an indium target with the γ-rays emitted by 0.5 g ^{226}Ra to obtain isomeric activity by the ^{115}In(γ, γ')^{115}In^m reaction. The negative result of their experiments is attributed by Guth [41Gu], [45Gu] to the fact that the monoenergetic γ-energies from ^{226}Ra do not correspond to the possible excitation levels of indium. Later, the possibility of performing such investigations was greatly influenced by artificially produced radioactive γ-sources. Intense radiation sources can be produced in nuclear reactors, with activities of the order of 100 TBq (~2700 Ci). Such a high primary (monoenergetic) γ-intensity produces a continuous spectrum – through Compton scattering – in the source and the intermediate material.

Harbottle [54Ha] measured ^{115}In and ^{111}Cd isomeric activities, following irradiation, with γ-rays of 3.7–66.7 TBq ^{60}Co and 48 TBq ^{182}Ta sources. Ikeda and Yoshihara [56Ik], [58Ik], [59Ik], [60Yo] measured the experimental cross-sections for ^{87}Sr^m, ^{103}Rh^m, ^{107,109}Ag^m, ^{111}Cd^m, ^{115}In^m and ^{176}Lu^m production and independently Veres [63Ve1, 2] also determined the integral and experimental cross-sections for the above-mentioned five (^{87}Sr^{m}, ^{103}Rh^{m}, ^{107,109}Ag^{m}, ^{111}Cd^{m}, ^{115}In^{m}) and for seven more (^{77}Se^{m}, ^{89}Y^{m}, ^{179}Hf^{m}, ^{191}Ir^{m}, ^{195}Pt^{m}, ^{197}Au^{m}, and ^{199}Hg^{m}) isomers. Under the given experimental conditions the number of Compton-scattered γ-quanta in the unit energy interval per primary γ-quantum (from the source at a threshold energy of about 1.1 MeV) was found to be 1.7×10^{-7} eV^{-1}. In view of this, the integrated cross-section values were found to be between 3 mb eV and 600 mb eV, respectively.

1.2. Conventions, definitions, units, and symbols

Conventions

Throughout this book, except where it is explicitly stated, absolute SI units of time, length, and mass have been used. However, for convenience, energies in eV, keV, or MeV and, usually, cross-sections measured in barns [1×10^{-28} m^2] have been given. Electromagnetic radiation is characterised usually by photon energy, except where explicitly noted, rather than by wavelength or frequency.

Definitions

a) *Activation levels (i.e. Al, As, gateway states):* The nucleus must have an E_a level that is greater than the energy E_m of the isomeric state. This level must have the appropriate spin-parity such that it has a high probability of being excited from the ground state and, at the same time, it has a large branching ratio for decay through states which feed the isomeric state.
b) *Primary γ-ray energy (E_s):* The E_s of the radioactive source should be equal or greater than that of the activation level, i.e. $E_s \geq E_a$. If the width of the radioactive source line E_s and that of the activation level E_a are such that there is little or no overlap of energy of the direct source spectrum with the resonance absorption line in the target nucleus, the geometry and distribution of the sources used to activate the target must be such that Compton-scattering takes place in the material surrounding the sources and target; this is in order to provide the necessary overlap. These conditions usually exist in intense radiation sources.
c) *Bremsstrahlung:* There are similar requirements for *bremsstrahlung* of an accelerator as in the former point; $E_x \geq E_a$.
d) *Cross-sections (σ, Σ or $\sigma\Gamma$):* In the case of a single Al σ, while Σ or $\sigma\Gamma$ are used, this occurs if the excitation can proceed via a group of Als.
e) *Effective cross-section (σ_{eff} or Σ_{eff}):* This is when the parameters of the activation levels (E_a, spin, parity, etc.) are not determined.

f) *Integrated cross-section:* This is when a single activation level (σ_{int}) is known and when a group of activation levels (Σ_{int} or $\sigma\Gamma$) in an energy range are presumed.

g) *Processes in the nucleosynthesis:* these include the *s-process, r*-process, *p*-process, *e*-process, or ν-process (to be explained shortly in section 4.2).

h) *The cross-section for a stimulated emission:* This is deduced from the Einstein formula: $\sigma_s = (\Lambda^2/2\pi)(\Gamma_r/\Gamma)$, where Γ_r is the radiation width and Λ is the wavelength of a transition between the two levels.

i) *Intermediate structure (IS):* This is a representation of the resonant photo-excitation mechanism, which is characterised by its excitation energy and its natural width.

j) *Total, partial and radiative widths* of an energy level are represented by (Γ, Γ_p, Γ_r), respectively.

k) *Number of density of nuclei in a given medium:* This is represented by N (g^{-1}, or cm^{-3} initial: N_i, final: N_f,)

l) *Graser* (Gamma-ray laser). Three features common to all lasers are (i) an active medium (e.g., atoms, ions and molecules); (ii) a host (this serves to confine the active medium to a definite volume and geometry; examples include CO_2 gas in a glass tube, Cr^{***} in a sapphire crystal, and Nd^{***} in a glass rod. Any graser would necessarily require a crystal host (cooled to the lowest possible temperature, given that a practicable recoilless emission seems to be an essential feature); and (iii) a pump (this supplies energy in order to excite the active medium to an inverted condition).

Symbols, dimensions, values, units and constants:
Some constants and units which are often used are presented in *Table 1.*

Table 1. Symbols, dimensions, and values of some units and constants used in this work

Unit of nuclear cross-section: barn, $1\ b = 10^{-28}$ [m^2]	Avogadro number: 6.0247×10^{23} [mol^{-1}] Phys. scale
Velocity of EM radiation: $c = 2.998\times10^8$ [$m\ s^{-1}$]	Boltzmann constant: $k = 8.617\times10^{-8}$ [$keV\ K^{-1}$]
Energy: 1 [keV] = 1.6×10^{-16} [J]	Mass of proton: $m_p = 1.67248\times10^{-27}$ kg
Gravitational acceleration at sea level: 9.806 [$m\ s^{-2}$]	Energy equivalents to m_p 9.315×10^5 keV
Planck constant, ratio of energy to frequency: $h = 4.136\times10^{-18}$ [keV s]	1 pc (parsec) = 2.8383×10^{16} m $\approx$ 3 light-years 1 light-year = 9.46×10^{15} m
σ_a Average cross-section of a medium for scattering or absorbing a photon [m^2, or $b = 10^{-28}\ m^2$]	Wavelength (Λ or λ) associated with 1 keV = 1.23978×10^{-9} m

This work has two main purposes: first, to collect the experimental and theoretical results of photoexcitation, to collate the efforts to develop stimulated emission devices for a graser pump, and then to apply the measured data in astrophysical situations; the second aim is to discuss or evaluate data or experiments involving the photoexcitation of isomers, and also cases in which applications of isomer activation have been used in the study of the structure of the nucleus. Discussion of some main equations and processes will be necessary; this may be in connection with the photoexcitation of isomers of stable nuclei, involving connections with nucleosynthesis and some other problems of the graser technique.

Before considering the above-mentioned problems a short enumeration of earlier applications will be given.

1.3. The applications of isomer activation

The photoactivation of isomers of stable nuclei has proved to be a useful tool for obtaining additional information on nuclear states. With the increasing interest in high activity sources for irradiation for practical purposes, information about the isomer activity produced is very important. The photoexcitation of some elements is a useful method in the field of activation analysis. The method, however, is between 2 and 4 orders of magnitude less sensitive than neutron activation analysis. On the other hand, it has the advantage that the samples do not contain too complicated spectra and it is a non-destructive method. The use of photoexcitation reactions seems to be a suitable method for measuring the intensity of high activity radiation sources as well.

- In the initial stage the photoexcitation of nuclear isomers was investigated for many elements found to have isomeric states in numerous laboratories (inclusive of practical applications, too). In 1980 Veres published a summary of "Gamma activation of nuclear isomers and its applications" in the *Atomic Energy Review* [80Ve1]; thus those applications are not discussed here. However, there is a short survey of these applications and the same references are given.
- Photoactivation analysis by the excitation of the isomers of stable nuclei having a half-life greater than 0.3 s has already been undertaken. Lukens and coworkers [61Lu], [71Lu], [61Ot], using a 3 MeV Van de Graaff accelerator and a 15 MeV linear accelerator, and Kaminishi and Kojima [63Ka], using a 6 MeV linear accelerator have examined most of the elements for isomerism. They have published tables of the isomers observed, their half-lives, transition energies, and yields. Additional information can be found in the references [61Yo], [63Bi], [63Ve1,2], [67Ab], [69La], [69Ve], [71Lu], [78Ve2].

– Determination has been made of the rare metal content in raw material [76Oc], [80St].
– Quantity determination has been carried out of certain elements in food and medicine [67Ab2], [77Sm], [77Ve], [78Hs], [78Pa], [90La].
– Concentrations of several elements have been determined in industrial sewage and fertilisers [77Ch].
– Continuous control of metal content – e.g., Cd – has taken place in the case of hazardous waste [51Sc], [77Ch].
– Non-destructive determination has been made with regard to noble elements (Ag, Au, Pt) in jewels, curio coins and numismatics [72Me2], [81Bo], [82Ve].
– The minimum detectable quantity of stable elements has been examined: [61Lu], [61Ot], [63Ka], [66Ve2], [69Ve], [71Lu], [79Br], [80Ve1,2].
– Activity determinations have taken place of strong γ-radiation sources, as well as the measurements of high dose rates [54Ha], [61Yo], [67Ve], [70La].
– Gamma activation has been used for the non-destructive assay of spent fuel assemblies [78Ge], [85La], [88Ve].
– Hot atom chemistry applications have been examined [77Yo], [88Ve].

This monograph has also been prepared with the aim of using the results of photoexcitation in two new branches of research: namely, the impact of (γ, γ') reactions on the evolution of the universe and in the technical development of graser techniques. It also deals with experimental techniques and theoretical methods developed in the field involving the excitation of stable nuclei by γ-quanta.

It is convenient to classify investigations into photoexcitation by the types (experimental or theoretical) of study or by the applied irradiation source (*bremsstrahlung* or monoenergetic source). For each stable nucleus having an isomeric state it is necessary to investigate the following: targets, irradiation facilities, experimental and theoretical evaluations, interaction mechanisms, the most probable excitation and partial γ-decay schemes. The cross-sections of the isomer activation, as well as other data about the nuclei have, as far as possible, been tabulated. A comparison has also been made between the data of the experimental and theoretical values of the integrated cross-sections, data for the excitation level energies and the life times of nuclei induced (by *bremsstrahlung*) from different accelerators, Compton-scattered monoenergetic radiation sources, positron radiation sources, and other irradiator facilities. A comparison has also been carried out of values determined by different mechanisms of the interaction: e.g., resonance and non-resonance fluorescence absorption, nuclear excitation by inelastic photoelectric effect, nuclear excitation by positron annihilation, and nuclear excitation by electron excitation (NEET).

2. The investigated isomers and applied irradiation facilities

Instead of providing an overview of the numerous studies which have been published over the last fifty years, here reference is made primarily to articles giving more general guidance as well as those necessary for supporting or emphasising specific points and providing details about particular milestones.

2.1. Isomers in the chart of nuclides

It is now believed that neutral atoms consist of N neutrons, Z protons, and Z orbital electrons, where $A = N + Z$. Isotopes are nuclides with the same Z but different N. It is also possible to classify the 266 stable nuclides by the evenness or unevenness of Z and N to give four possible categories. This can be advantageous in the choice of target material because, for example, isomers of even-even nuclides are very difficult to produce and, in most cases, they cannot be observed in (γ, γ') experiments. For instance, Kaminishi and Kojima [63Ka] – using a 6 MeV linear accelerator for irradiation – could not observe the isomer activities of even-even nuclides such as ^{90}Zr, ^{178}Hf, ^{190}Os, and ^{204}Pb in their experiments. Veres et al. [73Ve3] also irradiated these even-even nuclides and also ^{180}Hf with a very intense radiation source ($\sim 1 \times 10^{14}$ $\gamma\,s^{-1}$) of ^{24}Na having $E_{\gamma 1} = 1.38$ MeV, and $E_{\gamma 2} = 2.75$ MeV energies; no measurable isomeric activity of the samples was observed.

Table 2. Distribution of stable nuclides and their isomers

Proton (Z)	Neutron (N)	Mass (A)	Nuclides	Isomers
Even	Even	Even	159	8
Even	Odd	Odd	53	23
Odd	Even	Odd	50	10
Odd	Odd	Even	4	2
Total number of nuclides and isomers:			266	43

Table 2 shows that, for the odd-*A* nuclides, there are approximately as many nuclides with an even number of protons (even *Z*) as those with an even number of neutrons (even *N*). This evidence indicates that the nuclear force between two nucleons is independent of whether the nucleons are protons or neutrons. Uneven-uneven stable nuclides are scarce. Their scarcity is due to a *pairing energy* between particles in the same shell. The condition of being in the same shell increases the binding energy of these particles, making them more stable than particles in different shells. An odd-odd nuclide contains at least one unpaired proton and one unpaired neutron. These are usually in different shells and hence contribute weakly to the binding.

Table 3. The codes of shadows

Half-live of isomers (number)	shadow
0.3 seconds (s) to 1 hour (24 isomers)	10%
1 hour to 1 day (9 isomers)	20%
1 day to 1 year (10 isomers)	30%
>1 year (4 isomers)	40%

The shadow coding *(Table 3)* used in this edition of the chart of isomers *(Table 4)* emphasises the half-lives of isomers. Four different shadows – 10%, 20%, 30% and 40% – are used for the identifications of half-lives. In this instance the 10% shadow was chosen for the lowest half-life values, because these represented the highest value of isomerism, and the progression continued through 20% and 30%, with 40% being chosen for the highest half-life values.

Table 4 shows the isomers of stable nuclei which have longer half-life times than 0.3 of a second. Note that in certain cases in which the ground state data are displayed the $^{210}Bi^{m}$ isomer has a 3.04×10^{6} *y* half-life but the ground state is 5.013 *d*. Another one is $^{180}Ta^{m}$, with a half-life $>1.2 \times 10^{15}$ y. It has a 0.0012% abundance in nature and its ground state has a 8.152-hour half-life. Abrams et al. [66Ab] irradiated a germanium target with γ-rays from $^{116}In^{m}$ radiation sources ($E_{\gamma,\,\mathrm{max}}$ = 2.112 MeV) and observed no isomer activity.

In the vertical column on the left side of the map the squares contain the proton number (*Z*) of the nucleus, while the horizontal row on the top denotes the number of neutrons in the nucleus. Proton and neutron numbers are not strictly successive, because those having no isomeric states are omitted. An arbitrary cut-off point with a half-life of 0.3 s was chosen (for convenience) in presenting the data. For nuclei having more isomeric states only one is shown with a half-life showing lower energy.

Table 4. Chart of the isomers [a)] of stable [b)] isotopes, discussed in this monograph

N→	13	41	43	44	47	49	N→	50	52	56	58	60	62
Z↓ 13	$^{26}Al^{m}$ 6.5 s						Z↓ 39	$^{89}Y^{m}$ 16 s					
32		$^{73}Ge^{m}$ 0.5 s					40	$^{90}Zr^{m}$ 0.8 s					
34			$^{77}Se^{m}$ 17.4 s				41		$^{93}Nb^{m}$ 16 y				
35				$^{79}Br^{m}$ 4.9 s			43			$^{99}Tc^{m}$ 6 h			
36					$^{83}Kr^{m}$ 1.8 h		45				$^{103}Rh^{m}$ 56 m		
38						$^{87}Sr^{m}$ 2.8 h	47					$^{107}Ag^{m}$ 44.3 s	$^{109}Ag^{m}$ 39.6 s
N→	63	64	65	66	67	69	N→	71	73	75	77		
Z↓ 48	$^{111}Cd^{m}$ 49 m		$^{113}Cd^{m}$ 14 y				Z↓ 52	$^{123}Te^{m}$ 120 d	$^{125}Te^{m}$ 57.4 d				
49		$^{113}In^{m}$ 1.7 h		$^{115}In^{m}$ 4.5 h			54			$^{129}Xe^{m}$ 8.9 d	$^{131}Xe^{m}$ 12 d		
50					$^{117}Sn^{m}$ 13.6 d	$^{119}Sn^{m}$ 293 d							
N→	79	80	81	99	105	106	N→	105	106	107	108	109	
56	$^{135}Ba^{m}$ 1.2 d	$^{136}Ba^{m}$ 0.3 s	$^{137}Ba^{m}$ 2.6 m				72	$^{177}Hf^{m}$ 51 m	$^{178}Hf^{m}$ 31 y	$^{179}Hf^{m}$ 18.7 s	$^{180}Hf^{m}$ 5.5 h		
68				$^{167}Er^{m}$ 2.5 s			73			$^{180}Ta^{g}$ 8.1 d			
70						$^{176}Yb^{m}$ 11.4 s	74					$^{183}W^{m}$ 5.2 s	
71					$^{176}Lu^{m}$ 3.64 h								
N→	113	114	116	117	118	119	122	125	127	143			
76	$^{189}Os^{m}$ 5.8 h	$^{190}Os^{m}$ 10 m	$^{192}Os^{m}$ 6.1 s										
77		$^{191}Ir^{m}$ 4.9 s	$^{193}Ir^{m}$ 10 d										
78				$^{195}Pt^{m}$ 4 d									
79					$^{197}Au^{m}$ 7.73 s								
80						$^{199}Hg^{m}$ 43 m							
82							$^{204}Pb^{m}$ 1.12 h	$^{207}Pb^{m}$ 0.8 s					
83									$^{210}Bi^{g}$ 5 d				
92										$^{235}U^{m}$ 25 m			

a) The shadow coding used emphasises the half-lives of isomers for identification. 10% was chosen for the lowest half-life values, because this represented the highest value of isomerism and the progression continued through 20% and 30%; thus the 40% was chosen as the highest half-life value. A few nuclei had two or more isomeric states, but only one of their more significant isomers was displayed.

b) Three nuclei are absent from a terrestrial state due to the fact that their half-lives are 7.4×10^{5} y (^{26}Al), 2.1×10^{5} y (^{99}Tc) and 3×10^{6} y ($^{210}Bi^{m}$). The half-lives of $^{180}Ta^{g}$ and $^{180}Ta^{m}$ (as the rarest "stable" isotope) are 8.1 d and $\geq1.2\times10^{15}$ y, respectively.

2.2. Irradiation facilities

Among the many varieties of high-energy radiation sources, the particle accelerator possesses a unique combination of desirable characteristics, including:

- Homogeneous and controllable particle energy, in the 0.5–50 MeV energy range;
- Well-focused particle beams, and operation in either the direct current mode or pulsed mode;
- Freedom of particle choice for acceleration, including electrons, protons, deuterons, alphas, and heavy ions;
- Prolific beams of secondary particles (α, p, and n) and radiation, such as penetrating X-rays *(bremsstrahlung)* and controllable monoenergetic neutrons;
- Readily calibrated parameters of performance.

These accelerator systems are widely used in programmes involving precise nuclear research, photoexcitation, clinical therapy and industrial radiography, respectively. Over 500 Van de Graaff electron accelerators, betatron, dynamitron and other particle and heavy-ion accelerators (in the range of GeV) are in operation throughout the world. Most of these accelerators and X-ray generators are used for photoexcitation and produce particles and radiation in the 1 to 14 MeV range. Therefore only those laboratory facilities are mentioned which are used in isomer activation experiments.

An indium isomer was first produced by irradiation with X-rays of 1.85 MeV maximum energy from an impulse generator [39Po]. During the last 60 years more than 50 irradiation facilities have been used in photoexcitation experiments, among them a few γ-radiation sources (^{60}Co, ^{24}Na, ^{46}Sc, ^{142}Pr, and ^{137}Cs) and a 4 MeV LINAC in the Institute of Isotopes (Budapest). Much of the early systematic work in photonuclear activation analysis can be attributed to Engelmann [64En], and Lutz [71Lu]. Subsequently, articles in journals and conference proceedings numbering more than 300 presented the determination of about 40 of the elements with photons, varying in energy from 0.5 MeV to 50 MeV.

The other group of irradiation equipment used widely in isomer activation experiments includes radioactive sources emitting monoenergetic γ-rays. The main sources are ^{137}Cs, ^{46}Sc, ^{60}Co, ^{142}Pr, ^{182}Ta, ^{116}In^{m} (in a reactor loop), and ^{24}Na which has monoenergetic γ-rays from 0.662 MeV to 2.75 MeV. The positron sources used for photoexcitation are mainly ^{22}Na and ^{64}Cu.

Table 5. Typical characteristics of irradiation-facility accelerators, radioactive γ-sources, positron radiation, and neutron capture γ-lines used for photoactivation

No.	Type of equipment	Radiation (E_{max})	Intensity, beam-curve	Laboratory and Reference	
1	Impulse Generator	X; <1.85 MeV			[39Po]
2	ME[a)], ^{226}Ra, 0.5 g	γ, <2.45 MeV	(0.72) $1.3\times10^{10}\ \gamma\ s^{-1}$		[39Po]
3	Pres. El. Stat. Gen.	X, 1 to 3 MeV	100 μA at 0.9 E_{el}	Notre Dame,	[39Co]
4	Van de Graaff-Herb	X, <3.3 MeV		Notre Dame,	[45Wi]
5	ME[a)], ^{60}Co; ^{182}Ta	γ, <1.3 MeV	1.8; 1.3 kCi[c)]	BNL Uptown,	[54Ha]
6	ME[a)], ^{60}Co	γ, <1.33 MeV	37–74 TBq	Tokyo Univ.,	[58Ik]
7	Van de Graaff Gen.	X, E_{max}=15 MeV		Emerville, Calif.,	[61Lu]
8	Linear Accelerator	X, 2–6 MeV	20–100 μA (average)	GIRI, Nagoya,	[63Ka]
9	ME[a)], ^{60}Co	γ, <1.33 MeV	$3.2\times10^{13}\ \gamma\ s^{-1}$	IKI, Budapest,	[63Ve1,2]
10	Van de Graaff Gen.	e, 2.5–3.7MeV		MIT, 1940,	[65Ch]
11	ME[a)], ^{60}Co	γ, <1.33 MeV	$2.15\times10^{14}\ \gamma s^{-1}$	IKI, Budapest,	[66Ve4]
12	Van de Graaff Gen.	X, 3 MeV±10keV	$10^{10}\ \gamma$/MeV, 150 μA	Boston Univ.,	[67Bo]
13	ME[a)], ^{116}In^{m}	γ, <2.1 MeV	reactor-loop	IF, Riga,	[66Ab]
14	ME[a)], ^{24}Na	γ, <2.75 MeV	$2\times3.7\times10^{12}\ \gamma\ s^{-1}$	IF, Riga,	[69Ab]
15	ME[a)], (n,γ) reaction	γ, 5–9 MeV		Texas Univ.,	[69Ra]
16	ME[a)], ^{60}Co	γ, <1.33 MeV	$10^{13}\ \gamma\ cm^{-2}\ s^{-1}$	IKI, Budapest,	[69Ve]
17	ME[a)], ^{24}Na	γ, <2.75 MeV	$2\times1.7\times10^{12}\ \gamma\ s^{-1}$	IKI, Budapest,	[70La]
18	Electron accelerator	X, <5.6 MeV		Swarthmore,	[72Me2]
19	ME[a)], ^{22}Na	β^{+}, source	0.26 GBq	Kyoto Univ.,	[72Mu]
20	ME[a)], ^{24}Na	γ, <2.75 MeV	$2\times9.8\times10^{13}\ \gamma\ s^{-1}$	IKI, Budapest,	[73Ve3]
21	ME[a)], ^{46}Sc	γ, <1.12 MeV	$2\times1.7\times10^{13}\ \gamma\ s^{-1}$	IKI, Budapest,	[73Ve3]
22	ME[a)], ^{142}Pr	γ, <1.576 MeV	$3.1\times10^{11}\ \gamma\ s^{-1}$	IKI, Budapest,	[73Ve3]
23	ME[a)], ^{60}Co	γ, <1.33 MeV	$1.6\times10^{16}\ \gamma\ s^{-1}$	Tashkent,	[73Za]
24	ME[a)], (n,γ) reaction	γ-ray, 5–11 MeV		NRC, Soreq,	[70Mo], [74Ar]
25	ME[a)], ^{182}Ta	γ-ray, 1–1.3 MeV	$1.5\times10^{12}\ \gamma\ s^{-1}$	IKI, Budapest,	[78Ve2]
26	LINAC 7 MeV	X, 6–8, 4–16 MeV	650 mA, 2.1 μs	CNRS, Paris,	[79Br]
27	ME[a)], ^{64}Cu	β^{+}-source	51.8 GBq	Kiev, Nucl. Inst.,	[79Vi]
28	LINAC	8 MeV	Average curr.: 1 mA	Moscow, R. I.,	[81Bo]
29	Dynamitron	X, E_0= 4.3 MeV	$I_{el}\approx0.8$ mA	Stuttgart Univ.,	[89Pi]
30	Photon factory KEK	SR[b)], 40–357 keV	150 mA, storage ring	Tsukuba,	[89Yo]
31	Betatron	x, 3–4 MeV	2 μs, 5×10^{9} el./2 μs	Leningrad,	[90Po]
32	ELINAC	X, 14–30 MeV	100 μA	Sendai,	[91Sá]
33	Impulse LINAC	X, 4 MeV	25 μA	IKI, Budapest,	[91La2]
34	S-DANILAC	x, 2–7 MeV	Nom. curr.: 5 μA	T.H. Darmstadt,	[93Ca1]
35	M-30 Microtron	X, 4–15 MeV		I. Ya. I. Kiev,	[93Ma1]
36	ME[a)], ^{137}Cs	γ-ray, 661 keV	$6.3\times10^{13}\ \gamma\ s^{-1}$	IKI, Budapest,	[95Kh]
37	Roentgen generator	x, 200–300 keV		OMH, Budapest,	[95La3,4]

a) ME (monoenergetic radiation); b) SR (Synchrotron radiation); c) 1 Ci = 37 GBq

Table 5 presents a significant number of those facilities which have often been applied in the photoactivation of isomers.

Furthermore, it should be noted that recently an astronomical instrument for the observation of gamma-ray lines of the galaxy has been launched into space. The Arthur Holly Compton Gamma-Ray Observatory (Compton) was launched on 5 April 1991. The mission goal for the "Compton" is to perform broad-band γ-ray observations with better angular resolutions, a border range, and a greater degree of sensitivity than in previous missions. The "Compton" detects γ-rays [94Kn] with good angular resolution (about 1 degree) and an energy resolution of about 9% at 1 MeV. At the moment it is in the process of performing the first all-sky survey in an energy range from 1 to 30 MeV.

2.3. Isomer production measurement and evaluation of experimental data

The experimental procedure may be seen in *Fig. 2.* The first block represents the irradiation processes producing the isomer state of a target nucleus by *bremsstrahlung* of LINAC-s [60Wi], [62Bo], [62Ne], by Compton scattered radioactive γ-ray and by any other radiation source. The second block illustrates the activity measurement of isomers including the intensity and energy spectrum. The third block demonstrates a possible excitation and de-excitation mechanism of the nuclei after the irradiation by (γ, γ') reaction and also points to a possible stimulated process which may be fitting for pumping a graser.

In the third part *Fig. 2* the two arrows ↑ illustrate the transition probabilities for the excitation of a nucleus from the ground state to a single activation level (AL) and/or a group of intermediate (shaded) states (IS). From these (AL or IS) excited states the nucleus is partly de-excited by direct transition to the ground state and it also de-excites via direct or cascade transitions to the isomeric state. The arrow ↑ from the isomeric state represents a stimulation to an excited state and has a very short lifetime (*ns*, or *μs*). The dashed arrow illustrates the immediate de-excitation of the excited levels to the ground state which may serve to pump a graser.

Progress toward the graser goal is dependent upon research and technological development in many areas (see, for example, [81Ba]. This seems to urge further study of nuclear isomerism – e.g., level schemes, decay parameters, production reactions, and cross-sections for destruction – as well as for the production of the isomer by the pump. The central problem of γ-ray laser design is to pump the gadget without heating the lattice and so destroying the Mössbauer and Bormann (M&B) effects. Besides having a relatively low cross-section,

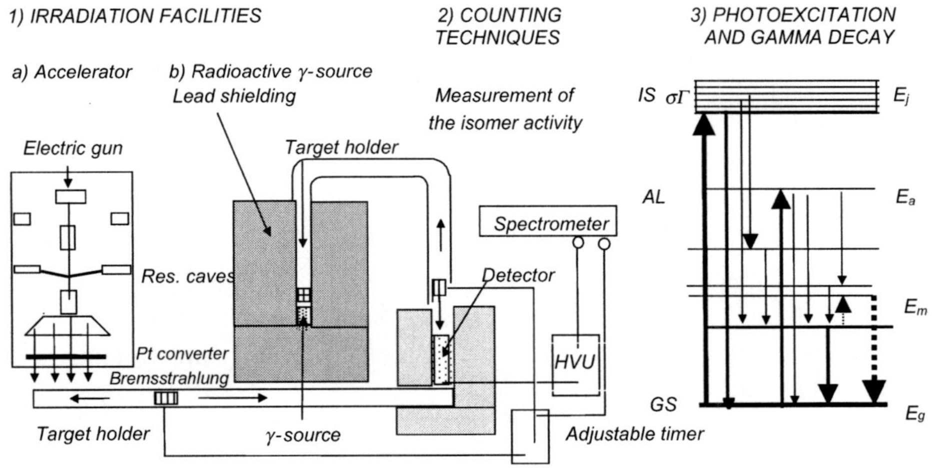

Fig. 2. A schematic drawing of the experimental process of photoexcitation with a theoretically possible γ-excitation and γ-decay scheme. The signs are: HVU (High-voltage unit, AL (Activation level), IS (Intermediate state), GS (Ground state). The ↑ arrow from the isomeric state represents a possible stimulation to an excited state and the dashed arrow represents the very quick de-excitation transition to the ground state

photons are intrinsically inefficient [98So]. Except for bound-bound X-radiation, all readily available sources of photons (at energies needed) are rather broad-band, calling for much higher power densities than the correspondent stored energy density (e.g., it would exceed 3×10^4 (J/cm^3). However, the intense heating by nuclear recoil and the ionisation that would be created within the graser solid by photoelectric and Compton processes are devastating to M&B effects.

3. The models and equations for the evaluation of photoexcitation data

During the evaluation of the measured data two well-known formulae were often used. In the first one Einstein predicted that induced transition must occur in quantum systems; this was based on a purely thermodynamic argument of general application. His derivation of

$$\frac{A}{B} = \frac{8\pi h}{\lambda^3}, \tag{1}$$

[the ratio of coefficients of probability for spontaneous (A) and induced (B) transitions between a pair of energy states] makes no reference to the type of system, other than that it can interact with an electromagnetic radiation of wavelength λ. This ratio, however, increases with the third power of the frequency (transition energy), so that spontaneous emission becomes an ever-stronger competitor to stimulated emission as the wavelength is reduced. The second one is the Planck formula:

$$u_\upsilon = \frac{8\pi\upsilon^2}{c^3} \frac{h\upsilon}{e^{\frac{h\upsilon}{kT}} - 1}, \tag{2}$$

where u_υ is the density of the energy, ν is the frequency, – c is the speed of light, $h\upsilon$ is the energy, k is the Boltzmann constant and T is the temperature.

The isomer production cross-section integrated over a single level or over a larger energy interval (where levels are overlapped) is:

$$\sigma_{\text{int}} = \int \sigma_{\text{iso}}(E)\, dE. \tag{3}$$

The rate of production of isomeric nuclei via a single excitation level by *bremsstrahlung* is approximately

$$Y = N(E_r)\, n\upsilon \int \sigma_{\text{iso}}\, dE, \tag{4}$$

where $N(E_r)$ is the incident flux at resonance, n is the number of nuclei per g, and υ is the thickness of the target in g/cm^2. This formula must be corrected for photon absorption and beam obliquity. For the continuum of (or densely lying) levels, one obtains:

$$Y = \int_{E_{\text{th}}}^{E_x} \sigma(E)\, f(E)\, dE. \qquad (5)$$

The photoexcitation of isomers was studied in different models, depending on the type of nucleus – for example, the odd-even nucleus was used to study the framework in the quasiparticle-phonon model (QPM), and in other types of nuclei applied in other models. Therefore, in this section a description is given of the theoretical framework of those models which were used for treaing such nuclei.

3.1. Nuclear excitation by electrons and *bremsstrahlung*

Isomer activation of nuclei by electrons and photons was the subject of investigation by researchers for over a decade during early experiments. The *bremsstrahlung* intensity was not known experimentally [54Mi], [56Bo] so that the $\int \sigma_{\text{iso}}\, dE$ could not be measured. Booth and Wright [62Bo] removed the difficulty of unknown incident flux intensity by measuring the resonance scattering of *bremsstrahlung* from nuclear states of known width. Chertok and Booth [65Ch] tested the cross-section for nuclear excitation by low (1–3 MeV) electrons, with measured lifetimes in ^{7}Li, ^{115}In and ^{111}Cd. The total cross-section for electron excitation of ^{115}In and ^{111}Cd was one order of magnitude larger than expected, near threshold. Agreement is obtained if the Robl cross-section is multiplied by the ratio of the Fermi function [$F(Z, p)$] and the forbidden β-spectrum shape ($S_L^{(L)}$) factor, $F(Z, p)/S_L^{(L)}F(Z, p_0)$. The combined results for the electron and photoexcitation are used to check the Robl theory.

a) Robl model for the calibration of nuclear excitation by electrons

The model chosen by Robl [56Ro] regards nuclear excitation by electrons as the inverse process of internal pair creation, where he assumed that both electron and nucleus can be regarded as point charges. The theory is valid only for low momentum transfer and the result explicitly contains the partial level width for gamma emission to the ground state. In the experiments the differential cross-sections ($d\sigma_{EL}$, $d\sigma_{ML}$) are measured and Γ_0 – the partial-level width – is known from resonance-fluorescence measurements. Therefore $d\sigma_{M1}$ can be compared with the experiments. The $d\sigma_{M1}$ can be integrated over an angle to give $\sigma(E) = g\Gamma_0(m_0c^2)^{-1}F_{\text{L}}$, where F_L depends on L (the angular momentum) and k (the energy of excited state in units of m_0c^2), respectively. F_L is given by Robl for all cases of L. When an isomeric state is excited via a higher state, the number of isomeric nuclei made per second by electron bombardment is

$$N_{e,\mathrm{iso}} = N_0 n\ dx\ \sigma_e(E)\Gamma_{\mathrm{iso}}/\Gamma, \tag{6}$$

while for photoexcitation

$$N_{p,\mathrm{iso}} = N(E_t)n\ dx \int \sigma_\gamma(E)\Gamma_{\mathrm{iso}}/\Gamma, \tag{7}$$

where N_0 and $N(E_r)$ are the number of electrons and protons (per MeV) incident per second on a target with $n\ dx$ nuclei per cm^2; $\sigma_e(E)$ and $\int \sigma_\gamma(E)\ dE$ are the cross-sections for electron- and photo-excitation of the higher state, and $\Gamma_{\mathrm{iso}}/\Gamma$ is the probability of decaying towards the isomeric state. Measurement of the electron activation of the isomer gives $F_L g\Gamma_0\Gamma_{\mathrm{iso}}/\Gamma$ and measurement of the photon activation gives $g\Gamma_0\Gamma_{\mathrm{iso}}/\Gamma$; thus the value of F predicted by the Robl calculation can be obtained. An independent measurement of $g\Gamma_0$ by the resonance absorption of photons gives a limitation to the possible values of L and determines $\Gamma_{\mathrm{iso}}/\Gamma$.

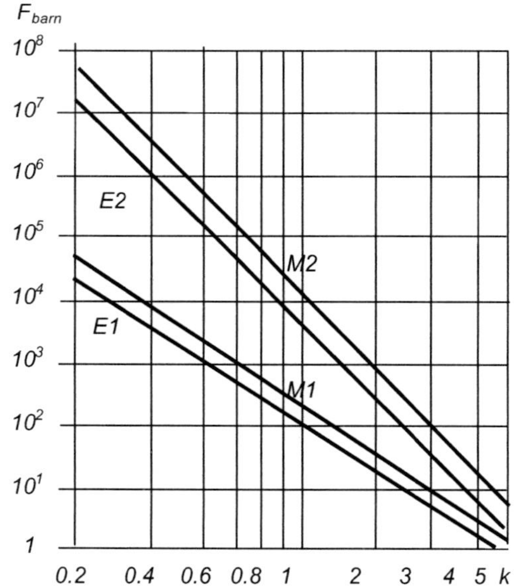

Fig. 3. The total cross-sections for an electron energy of 10 mc^2 and for excitation energies between 0.2 and 5 mc^2. The cross-sections are given for electric and magnetic multiple transitions in an arbitrary order. The F_E electric, and F_M magnetic values are the so-called Robl factors

The calculation of the cross-sections does not make explicit use of nuclear matrix elements, but involves only the mean lifetimes of the excited states. The cross-sections are given for electric and magnetic multiple transitions of an arbitrary order, and are shown in *Fig. 3.*

b) Calibration of pulsed bremsstrahlung *spectra*

As already mentioned, the (γ, γ') reactions for the production of isomers were first investigated in 1939 by *bremsstrahlung* [39Po]. However, after more than 50 years the difficulties arising with this type of work have not allowed experimental reproducibility to develop to the level expected with other types of reaction. Not more than forty or so papers are available in the literature (reference [86Yo] contains a thorough review) and the results reported for the same reactions show extreme variances. The difficulties in conducting (γ, γ') reactions are probably typified best by the excitation of $^{115}In^{m}$ ($t_{1/2}$ = 4.5 h, $E_{m\gamma}$ = 336 keV). The principal activation level that is radiatively connected to both the initial and final isomeric state lies at 1078 keV and can be conveniently excited by resonant absorption of Compton scattered gammas from a ^{60}Co source mounted in close proximity. Nevertheless, an experiment in 1981 produced such an excess in the isomeric population that it was raised to postulate a new mechanism in the ^{115}In for non-resonant absorption [81Lj]. However, this has been widely debated since then. For example, Yoshihara et al. [86Yo], using hot atom chemical studies (*tris*actylacenato indiumIII target), indicated that the non-resonant process must be less than 3.3% of the non-resonant effect.

Anderson and Collins [87An] published a method for the calibration of pulsed X-ray spectra in the 0.2–1 MeV energy range, to resolve the mentioned discrepancies. They applied a method for the calibration of ^{77}Se and ^{79}Br spectra produced by photoexcitation. The levels 249.7, 439.4 and 520.6 keV – from where there are direct transitions to the ground state and direct and cascade transitions to the isomeric state – were established to be activation levels.

The calibration of *bremsstrahlung* spectrum required some initial assumptions about its functional form.

- The high energy portion of the curve is a linear Kulenkampff approximation (see ref. [67Ku]) with endpoint energy E_m.
- At energy E_1 this joins a region of constant intensity which extends down to a cut-off level E_c. Thus the model can be expressed by $\phi(E) = \phi_0 (E_m - E) / (E_m - E_1)$, $E_1 \leq E \leq E_m$ and $\phi(E) = \phi_0$, $E_c \leq E \leq E_1$. The cut-off level E_c was chosen to be 200 keV since all of the pump lines lay above this value.

The model is illustrated in *Fig. 4.*

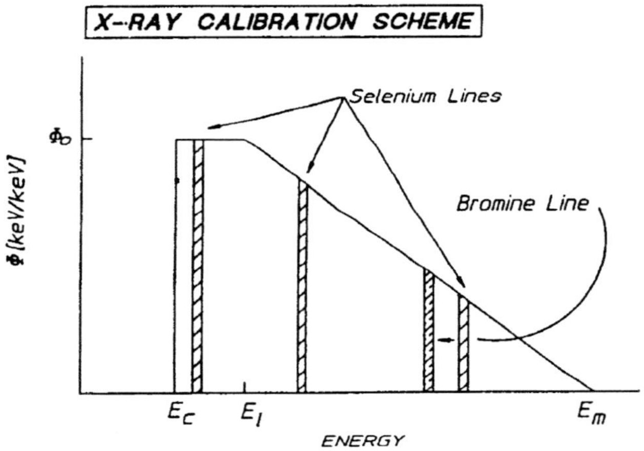

Fig. 4. Model X-ray spectrum used to calibrate the *bremsstrahlung* spectrum output. The respective locations of the Se and Br lines are shown. Most of the Se contribution occurs at the 480 keV line (which is actually two lines at 420 and 521 keV), and all of the Br contribution occurs at 761 keV from ref. [87An]

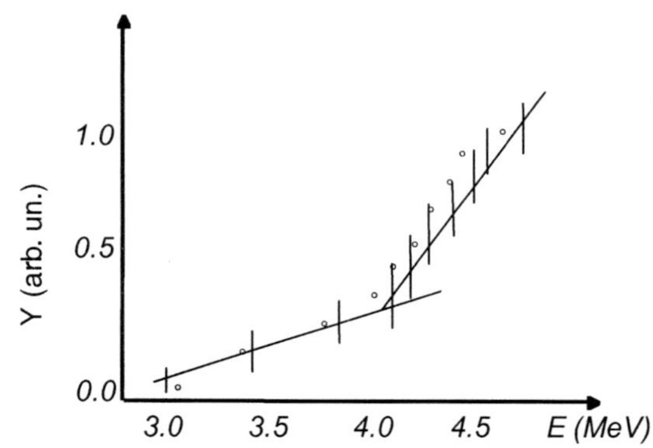

Fig. 5. The yield of the isomeric state in ^{81}Br as a function of the endpoint energy *E*, in arbitrary units

In their next paper [88An] the authors made an extension to a larger range (0.2–1.5 MeV) of photon energy and to intensities from 10^{12} keV/keV to 10^{16} keV/keV using the target nuclei ^{77}Se and ^{79}Br.

Ponomarev et al. [90Po] investigated the anomaly in the energy dependence of the yield of the isomerism in ^{81}Br by photoexcitation, regarded as a manifestation of a nuclear-intermediate model. An anomaly in the *bremsstrahlung* produced yield of the isomer with $9/2^+$, $E_x = 536$ keV; $t_{1/2} = 36$ μs is observed as a function of the endpoint energy in the energy range of 3–4 MeV. The isomeric yield was normalised by the fixed number of the accelerated electrons stopped in the betatron target by a tantalum sheet 1.5 mm thick – i.e. slightly thicker than the range of 4 MeV electrons. The isomer yield is presented in *Fig. 5.*

The activation level with E_{ex} = 3.18 MeV($5/2^+$) and with a wave-function with a 13% contribution of the quasiparticle component $2d_{5/2,}$ and a 83% contribution of the quasi-particle⊗phonon component $[1g_{9/2}2_1^+]$, which provides the resonance in the total scattering cross-section and is well linked to the isomeric state, is obtained in the framework of the quasi-particle phonon model. From the experiment it was concluded that the first component is the low-lying fragment of the quasiparticle state $2d_{5/2}$. The valence transition of the proton from the ground state is the reason for the strong radiatve excitation of this level; with a value $B(E1\uparrow) = 0.03\ e^2\ fm^2$. It would be interesting to obtain information from the direct excitation of the level with $J^\pi = 5/2^+$, E_{ex} = 3.45±0.15 MeV and $\Gamma_0 \sim 0.7$ eV in the reactions of elastic photon scattering with a low momentum of transfer.

3.2. Nuclear excitation by monoenergetic *γ-rays*

The monoenergetic γ-rays of radioactive sources have discrete energy and it is well known that a nucleus could not be excited directly to an isomeric state. Gamma-rays of energies matching the excitation energy levels of a nucleus within the region of the level width are resonantly absorbed with a very high cross-section. From the excited state the nucleus can decay to the ground state via other lower levels, either directly or via the isomeric level. The condition for isomer activation is that the primary γ-ray energy of the radioactive sources E_s must be equal to or greater than that of the activation level – i.e. $E_s \geq E_a$.

It can be seen from the above arguments that the spectral flux density is a result from the Compton scattering in the radiation source, intermediate material, and the target. The calculation of the resonance flux also plays an important role in the experimental determination of the partial-level width and integral cross-section. This quantity can influence very much the conclusions drawn from the experiments. For practical purposes we give here those approximate formulae from which the resonance flux at the target can be calculated taking into account the scattering in the radiation source, the intermediate material, and the target. The contribution from the scattering in the shield is omitted because it is negligible, since it is within the error limit. The deduction of the formulae and other features is also omitted because details can be found in Refs [60Yo], [66Ve4], [82Wa].

The experimentally determined cross-section is:

$$\sigma_{\exp} = \frac{I_k A(1+\alpha)}{\varepsilon \Omega S N \Phi a g_r}, \tag{8}$$

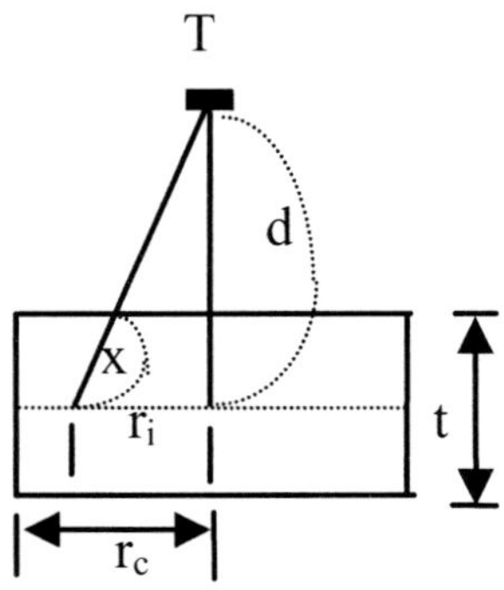

Fig. 6. Geometrical variables for the calculation of scattered spectra from disc source

where I_k is the initial number of isomeric decay per second, A is the mass number of the target, α is the internal conversion coefficient, $\varepsilon\Omega$ the geometrical coefficient, S the saturation factor, N the Avogadro number, Φ the γ-flux (cm^{-2} s^{-1}) of the sources, a the abundance of the isomeric nucleus, and g_r the weight of the target (in grams).

With the experimental conditions having been taken into consideration, the total spectral flux for disc and cylindrical sources are:

Scattering from the energy $h\nu$ to the $h\nu'+\Delta\nu'$ resonance range is calculated by the Klein–Nishina formula ($\Delta\Phi\nu$) valid for Compton scattering on atomic electrons which are free in this energy range.

$$\frac{\Delta\Phi}{\Delta h\nu'} = \frac{r_e^2 mc^2}{(h\nu)^2}\left[\frac{h\nu}{h\nu'} + \frac{h\nu'}{h\nu} - 2\left(\frac{mc^2}{h\nu'} - \frac{mc^2}{h\nu}\right) + \left(\frac{mc^2}{h\nu'} - \frac{mc^2}{h\nu}\right)\right], \tag{9}$$

where the classical electron radius $r_e = 2.81785\times10^{-13}$ cm; the rest mass of electron $mc^2 = 0.511$ MeV.

– **Disc** – from *Fig. 6* it can be seen that, in a disc-shape source in the T (target) point, the scattered different flux (F_d) is:

$$F_d = \frac{\Delta\Phi_e}{\Delta h\nu'} F_p \left\{ \frac{n_{01}}{\mu_i} \sum \exp\left(-\mu_1 x_i\right)\left[1 - \exp\left(\mu_1 x_i\right)\right] \frac{r_i \Delta r}{r_i^{\,2} + d^2} + \right.$$
$$+ \frac{n_{02}}{\mu_2{}' - \mu_2} \sum \exp\left(-\mu_2{}' x_i{}'\right)\left[1 - \exp\left(\mu_2 - \mu_2{}'\right) x_i{}'\right] \frac{r_i \Delta r}{r_i^{\,2} + d^2} + \tag{10}$$
$$\left. + \frac{n_{04}}{\mu_4}\left[1 - \frac{1}{\mu_4 l} + \frac{\exp\left(-\mu_4 l\right)}{\mu_4 l}\right]\right\}.$$

– **Cylindrical** – a typical arrangement of cylindrical shape of a source during the irradiation is shown in *Fig. 7* and the integral scattered flux can be calculated as follows:

$$F_c = \frac{\Delta\Phi_e}{\Delta h\nu} F_p \left\{ \frac{n_{01}}{\mu_1} \sum_{i=1}^{m} \exp\left(-\mu_1' y_i\right)\left[1-\exp\left(-\mu_1 y_i \frac{z_i \Delta z}{d^2+(z-h)^2}\right)\right]+\right.$$

$$+\frac{n_{02}}{\mu_2'-\mu_2} \sum \exp\left(-\mu_2 y_i'\right)\left[1-\exp\left(\mu_2-\mu_2'\right) y_i'\right] \frac{z_i \Delta z}{d^2+(z_i-h)^2}+ \tag{11}$$

$$\left.+\frac{n_{04}}{\mu_4}\left[1-\frac{1}{\mu_4 l}+\frac{\exp\left(-\mu_4 l\right)}{\mu_4 l}\right]\right\},$$

where $x = \frac{d}{2}\frac{\left(h^2+r^2\right)^{1/2}}{h} - t/2$ and $y_i = \frac{[d_2+(z_i-h)^2]^{1/2}}{d} - r_k$.

The remaining symbols in equations (8–10) are as follows: F_p is the flux of the primary γ-quanta; n_{01}, n_{02}, n_{04} are the electron density (cm^{-3}) of the radiation sources, the intermediate substance and the target, respectively; μ_1, μ_2, μ_4, and $\mu`_1$, $\mu`_2$ are the attenuation coefficients (cm^{-1}) of the same materials at the primary and resonance energy respectively; x_i, y_i, z_i and l are the thickness of the intermediate substance and target, respectively.

The summations in equations (10) and (11) are the approximations of the elliptical integral equations. This means that the radius of disc, and the length of cylindrical sources are divided into m pieces and, with a weighting factor for the absorption and the distance, an average is calculated. In the case of a radiation

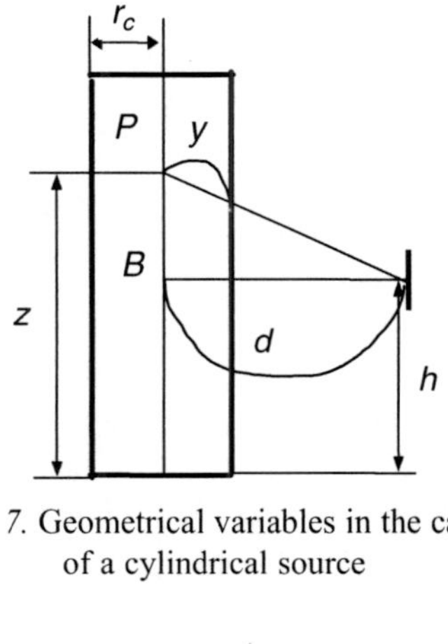

Fig. 7. Geometrical variables in the case of a cylindrical source

source having more γ-rays the formulae should be summed for all γ-rays to a given activation level. When there are more activation levels the number of resonance γ-rays should also be summed in line with those activation levels.

In the literature, calculations are given for single scattering events. Multiple scattering may be neglected if the scatterer is not very thick, but it cannot be neglected in the case of a thick scatterer nor if the energy of the primary γ-line is much higher than the energy of the activation level. In this latter case, for example, the number of single scattered γ-quanta should be 4–12% of the number of primary γ-quanta in a ^{24}Na source within a 1–10 cm radius and in the energy region 1–2.75 MeV. The γ-ray energy spectrum, caused by the scattering under spherical source conditions, is completely the same for any direction. However, for disc and cylindrical sources under certain conditions the effect of the angular distribution should also be calculated [79Wa2].

3.3. Resonance fluorescence measurements and essential equations

The theory and method of resonance fluorescence measurements have been reviewed by Metzger [59Me]. The cross-section for resonance absorption of unpolarized radiation is given by the Breit–Wigner formula [36Br]:

$$\sigma(E)=\frac{\lambda^2}{8\pi}g\frac{\Gamma_0\Gamma_m}{(E-E_r)^2+\frac{1}{4}\Gamma_t^2}, \tag{12}$$

where $g=(2J_r+1)/(2J_0+1)$, Γ_t is the total width of the excited state, J_r is the angular momentum of the excited state, J_0 is the angular momentum of the ground state, E_r is the resonance energy, $\lambda=hc/E_r$ and Γ_0 is the partial width of the excited state for photon decay to the ground state.

Integration of the Breit–Wigner formula for photon scattering gives:

$$\int\sigma(E)\,dE=\frac{1}{4}g\lambda^2\frac{\Gamma_0^2}{\Gamma}. \tag{13}$$

The sum of the partial widths to the ground state that is compatible with the conservation of angular momentum and parity is Γ_0. For example, in these cases of the transition of mixed $M1$ and $E2$ multipolarity, $\Gamma_0=\Gamma_0(M1)+\Gamma_0(E2)$ with the mixing ratio given by $\delta^2=\Gamma_0(E2)/\Gamma_0(M1)$.

The attenuation of *bremsstrahlung* intensity due to resonance absorption in a thin target is given by $\exp(-0.693\ \kappa t)$, where $\kappa=ng\lambda^2\Gamma_0/(4\pi^{1/2}\Delta)$, the

Doppler width Δ is given by $\Delta = E_r(2kT/Mc^2)^{1/2}$, T is the effective temperature of the compound, M the mass of the resonant nucleus, n the number of resonant nuclei/g and t the thickness (g/cm^2) of the target.

The effective temperatures used in some studies were obtained from a numerical integration involving the Debye temperature and the ambient temperature according to a formula given by Lamb [39La]. The scattered intensity from a thick scatterer can be calculated by an iterative procedure.

The scattering measurement determines $gW(\theta)\Gamma_0(\Gamma_0/\Gamma)$, where $W(\theta)$ is the angular distribution of scattered photons, and is given by $W(\theta) = 1+A_2P_2(\cos\theta) + A_4P_4(\cos\theta)$.

Here A_2 and A_4 are functions of the angular momentum of the ground and excited states, respectively, as well as the multipolarity of the transition $J_r \rightarrow J_0$ [57Bi]. The self-absorption measurements determine $g\Gamma_0$. Through a combination of the results of both methods, in some cases it is possible to deduce Γ_0/Γ (provided the angular distribution of the transition is known).

According to the above considerations, the experimental cross-section σ_{exp} can be related to the integral cross-section:

$$\sigma_{int} = \sigma_{exp} f_r / f_p \quad (cm^2\ eV), \tag{14}$$

where f_r is the number of γ-quanta scattered over the 1 eV energy region at the resonance energy. It also follows from the equations that it is possible to determine the half-life of the activation level with knowledge of the quantum numbers (i.e. spin and parity) and the branching ratios, B_0, B_i, as shown below:

$$t_{1/2} = gh\lambda^2 B_0 B_i / 4\sigma_{int}\,. \tag{15}$$

3.4. Spin-dependent gamma-ray cascade calculations

Inelastic gamma-scattering can add an important contribution to the information provided by nucleon emission studies of the giant dipole resonance. Since this energy region lies above the nucleon separation energy, the dominant decay mode is particle emission. The excitation function of some (γ, γ') reactions has been studied in some of the early works [51Sc], [56Me], [60Bo]. At the beginning of the 1990s the measurements did not confirm the earlier published structures in the case of $^{99}Tc^m$, $^{103}Rh^m$, and $^{115}In^m$. In order to clarify the situation activation experiments were also performed in this energy range (14–30 MeV) [91Se], [91Sá].

Due to the essential contribution of the pre-equilibrium process to the high-energy part of the γ-ray spectrum, however, the de-excitation through a gamma cascade only emphasises the de-excitation component. In addition, isomer-ratio calculations should involve angular momentum dependence of the branching ratios. Here a brief presentation is given of the γ-ray cascade calculations; this is similar to that of Pönitz [66Pö], but with corrected transition probabilities and open nucleon emission channels.

The photon-absorption by a zero-particle, zero-hole (0p0h) ground state was assumed to excite 1p1h configurations. Equilibrium level densities were calculated according to the parity-independent formula as:

$$\rho(E,J)=\frac{\left(\sqrt{\pi}\right)\exp\left[2\sqrt{a\,(E-\Delta)^2}\right]}{12a^{1/4}\,(E-\Delta)^{5/4}}R\,(J,\sigma), \tag{16}$$

where

$$R\,(J,\sigma)=\frac{(2L+1)}{(2\sqrt{2\pi})\sigma^3}\exp\left[-\frac{(2J+1)^2}{8\sigma^2}\right]. \tag{17}$$

E is the excitation energy, J is the spin, Δ is the pairing energy taken from [62Ne], a is the level density, and σ is the cut-off parameter. According to Williams [71Wi], the density of the quasi-bound particle-hole state ω was taken with an energy-dependent part and re-normalised following Akkermans and Guppelaar [85Ak].

The transition probability W from the initial state ($E,J \rightarrow U,S$) was assumed to have the form:

$$W\,(E,J\rightarrow U,S)=\left[\prod_{k=2}^{r}\frac{\langle\Gamma'_{kJ}\rangle}{\langle\Gamma_{kJ}\rangle}\right]\frac{d\Gamma^{\gamma,S}_{\text{equ},J}-dE_\gamma}{\langle\Gamma_{\text{equ},J}\rangle}+ \\ +\sum_{\Delta n=2,\,n=2}^{r-1}\left[\prod_{k=2}^{n-2}\frac{\langle\Gamma^{\downarrow}_{kJ}\rangle}{\Gamma_{kJ}}\right]\frac{d\Gamma^{\gamma,S}_{nJ}-dE_\gamma}{\langle\Gamma_{nJ}\rangle}. \tag{18}$$

The equilibrium differential γ-escape width is

$$\frac{d\Gamma^{\gamma,S}_{\text{equ},J}}{dE_\gamma}=\frac{\rho(U,\,S,-\pi)+\sum_{i=0}^{N}\delta(U-E_i)\delta_{JJi}\,\delta_{-\pi\pi i}}{\rho(E,\,S,\pi)}xE_\gamma^3\,f_{E1}\left(E_\gamma\right), \tag{19}$$

where $E_\gamma = E - U$ is the energy of the emitted γ-ray, f_{E1} is the γ-ray strength function and E_i, J_i, and π_i are the energy, spin, and parity values of the ith level respectively, in the region of discrete levels. In accordance with the Brink–Axel hypothesis, $f(E1)$ was taken in the form:

$$f(E1) = 4/3(hc)^2 \Sigma \sigma_i \Gamma_i^2 E_\gamma / [(E_\gamma^2 - E_i^2)^2 + E_\gamma^2 \Gamma_i^2], \quad (20)$$

where σ_i, Γ_i, and E_i are the peak height, width, and energy parameters of the photonuclear giant dipole resonance. $\Gamma_{equ.,J}$ is the sum of the equilibrium neutron, proton and total γ-ray widths. In the equilibrium state the neutron and proton emissions were calculated in the framework of the conventional statistical model. The average total widths $<\Gamma_{kJ}>$ for the k excitation state is given by $<\Gamma_{kJ}> = <\Gamma^{\downarrow}_{kJ}> + <\Gamma_{n,kJ}> + <\Gamma_{p,kJ}> + <\Gamma_{\gamma,kJ}>$, where $<\Gamma^{\downarrow}_{kJ}>$ is the average damping width, $<\Gamma_{n,kJ}>$ the average neutron width, $<\Gamma_{p,kJ}>$ the average proton width, and $<\Gamma_{\gamma,kJ}>$ is the total average pre-equilibrium differential γ-ray escape width.

3.5. The quasi-particle-phonon nuclear model

The first theoretical papers devoted to the damping of deeply bound hole states were published in 1977. The main ideas of the model were formulated in the early 1970s by Soloviev. In the QPM (quasi-particle-phonon model), as in many other models, the nucleus is considered as a system of interacting nucleons moving in a common average field. The interaction between them differs from those of free protons or neutrons and is chosen in a phenomenological way. The parameters of the effective interaction are fitted to the experimental data.

The problem of the interaction of an odd particle with a doubly-even core in odd-mass nuclei has attracted interest in the isomer activation experiments, too. At higher excitation energies a large number of other nuclear modes influence the damping of the motion of the odd particle. In nuclear transfer reactions these states manifest themselves as broad resonance-like structures (often called "bumps") superimposed in a large continuum. The bumps are very similar to the giant resonance observed in doubly-even nuclei by means of inelastic scattering of electrons and hadrons. The bumps contain a great deal of information. To describe the width of the single-particle state, one needs to calculate the mixing of the simple excitation mode with the states of the next level of complexity. This width is referred to as the "spreading" width and is denoted by $\Gamma^{\downarrow}$. Another width, called the "escape" width $\Gamma^{\uparrow}$, is associated with decay by particle emission. In heavy nuclei the escape width is only a small fraction of the observed width (i.e., $\Gamma^{\downarrow} >> \Gamma^{\uparrow}$). A simple estimation, based on the so-called golden rule, is $\Gamma^{\downarrow} = 2\pi V^2 \rho$, where V^2 is the average squared matrix element and ρ is the density of the excited states. In heavy nuclei the density of three- and five-quasi-particle states in the energy region of 10–20 MeV is of the order of 10^3–10^4 states/MeV.

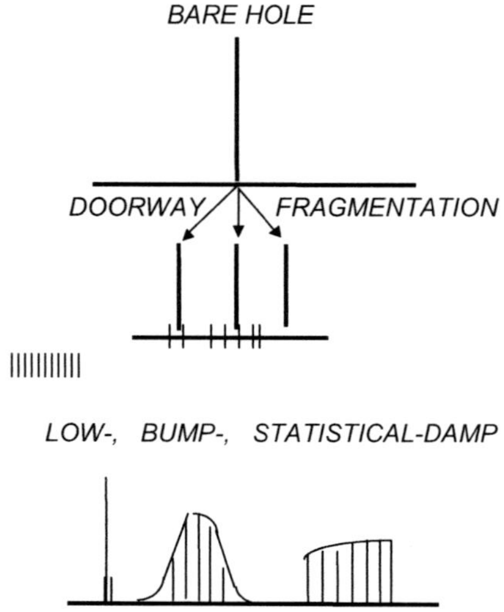

Fig. 8. The two stages of fragmentation of the single-particle state. The upper horizontal line represents the first step, when the bare single particle state is spread over several doorway states. In a second step the doorway states have their strengths dumped over the many other degrees of freedom, so the statistical damping strongly depends on the density level of the adjacent complicated states

The spreading is believed to occur in two stages, as is shown in *Fig. 8.* In the first stage of fragmentation, the bare single-particle state is spread over several doorway states. In a second step the doorway states themselves have their strengths damped over many other degrees of freedom. The amount of statistical damping depends strongly on the density level of adjacent complicated states. A useful tool for the description of the spreading width is the method of the strength function. This method gives information about the average values of observable quantities – i.e., centred energy E_x, the width $\Gamma^{\downarrow}$ and the total amount of strength ΣC^2S. At the level of high excitation energy only average values of this type have a physical meaning. They contain all the necessary information about the individual states and are not very sensitive to small changes of the model parameter. The experimental data are also presented as strength functions because only the average characteristic of the strength distribution (E_x, $\Gamma^{\downarrow}$, ΣC^2S) can be meaningfully compared to the predictions of the nuclear models. Numerous calculations connected with the damping of single particle modes in nuclei have been done within the quasi-particle-phonon model.

Gales et al. [88Ga2] presented a review of this model connected with the description of odd spherical nuclei which lacked *detailed* description.

General formalism

They supposed that the effective Hamiltonian of the QPM consists of four main parts: $H = H_{sp} + H_{pair} + H_M + H_{SM}$. Here H_{sp} describes the motion of independent nucleons in a self-consistent average field; H_{pair} represents the monopole pairing interaction in the particle-particle channel; H_M is the separable spin-multiple interaction in the particle-hole channel; H_{SM} is the separable spin-multiple interaction in the particle-hole channel.

The wave function and diagonalization

Since Gales et al. obtained the expression for the quasi-particle-phonon Hamiltonian, they were able to calculate the spectrum of excited states in odd-mass nuclei, and determined the wave function of an odd mass-spherical nucleus, which has the form:

$$\Psi_\nu(JM) = C_{J\nu}\left(\alpha^+_{JM} + \sum_{\lambda ij} D_j^{\lambda i}(J\nu)\left[\alpha^+_{jm}\, Q^+_{\lambda\mu i}\right]_{JM} + \right.$$

$$\left. + \sum_{\lambda_1 i_1 \lambda_2 i_2 jI} F_{jI}^{\lambda_1 i_1 \lambda_2 i_2}(J\nu)\left[\alpha^+_{jm}\left[Q^+_{\lambda_1\mu_1 i_1} Q^+_{\lambda_2\mu_2 i_2}\right] IM_1\right]_{JM}\right). \tag{21}$$

where α^+_{JM} and aJM are the creation and annihilation operators of nucleons having shell model quantum numbers njlm and energy E_j. The $Q^+_{\lambda\mu i}$ and $Q_{\lambda\mu i}$ are the creation and annihilation operators of phonons, λ is the chemical potential, $C_{J\nu}$ is the energy gap of the correlation function, $D(J\nu)$ is the diagonal matrix and $F(J\nu)$ is the transformation function. When only the "quasi-particle ⊗ two phonons" components are included in the wave function (21), the form is very simple, because the second term vanishes in the calculation. It follows from the calculations of the density of states that at excitation energies 7–8 MeV the greater part of the states have structures of the type "quasi-particle ⊗ three phonons" and "quasi-particle ⊗ four phonons". However, here interest is focused upon the distribution of the single-quasi-particle component, i.e. the coefficient $C^2_{J\nu}$. In addition, the operator $H_M{}^{qph}$ has the particular property of being able to mix states differing by one photon. This leads to the conclusion that components which have an influence will not change significantly. Nevertheless, it is clear that the wave function does not describe the structure of "real" states of the nucleus at an energy $E_x \sim 5$–10 MeV. It is also clear that the calculated values of $C^2_{J\nu}$ are equivalent to the single-quasi-particle strength in an energy interval which is roughly equal to the distance between states of the type "quasi-particle ⊗ two phonons". More detailed calculations and conclusions may be found in the above-mentioned [88Ga2] paper.

3.6. Higher-order processes in the photon nucleus interaction

Some discrepancies between theory and practice in experiments in photon-nucleus interactions, and in nuclear reactions induced by positrons are reviewed in the literature. The problems associated with reconciling the experimental results with theory have inspired several predictions of new higher-order processes in photon-nucleus interactions. After 18 years of research no satisfactory explanation had been found for the mechanism for non-resonant contributions. One approach assumed a nuclear Raman-scattering mechanism. In principle, all possible nuclear intermediate states should be considered but it was assumed that the low energy tail of the giant dipole resonance could serve as the intermediate states. The calculated value was many orders of magnitude below the experimental results. The possibility of nuclear levels being excited in positron annihilation was first discussed by Present and Chen [52Pr]. Subsequent experimental investigations have measured cross-section orders of magnitude larger than the theoretical predictions [72Mu], [73Vey], [79Ba2], [79Wa2], [80Wa], [81Lj], [82Kr], [82Pi], [85Pi], [87Pi].

It seems clear that there are still major difficulties in explaining low energy photon-nuclear interactions and positron-induced excitations.

Inelastic photoelectric effect (IPE)

Pisk et al. [82Pi] considered the possibility of an inelastic photoelectric effect. Feynman diagrams for this process are given at the top of *Fig. 9,* parts (a) and (b). In this third order process nuclear excitations occur in an interaction involving an incident photon, a bound electron, and the nucleus.

The process has obvious similarities with the internal Compton effect and Coulomb excitation. The electron-photon interaction can be treated in first order perturbation theory as the photoelectric effect, and the interaction between the electron and the nucleus can be treated by second order perturbation theory as the exchange of a virtual photon. Such an approximation led the authors [82Pi] to expect the accuracy of the calculations to be better for rather large photon energies, low Z values and for relatively low-energy nuclear states.

Nuclear excitation during pair production

Horvath et al. [84Ho] have studied a process which involves electron-positron creation in a nuclear Coulomb field by an incident photon, with the simultaneous excitation of the nuclear level. Feynman diagrams of this process are shown in the middle part of *Fig. 9* (a) and (b).

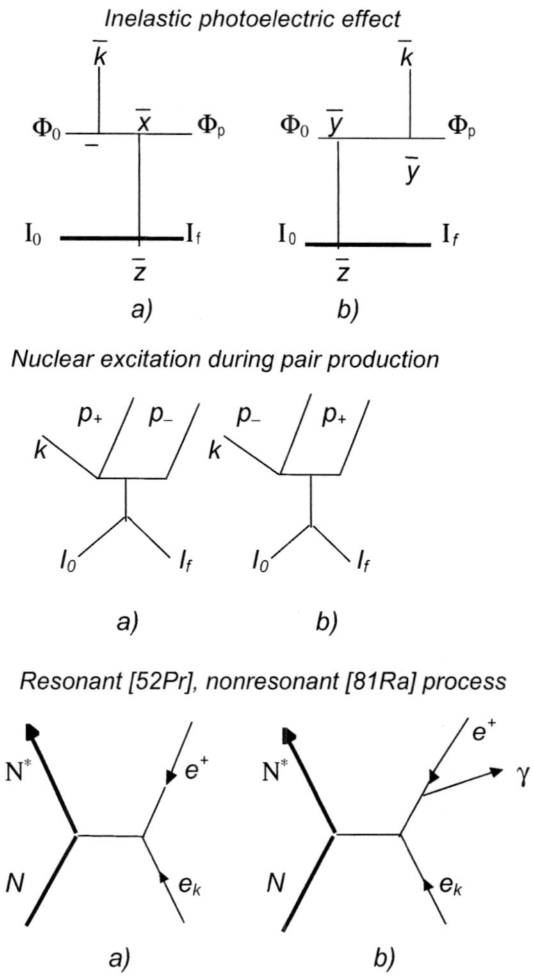

Fig. 9. Diagrams of higher-order processes for (a) inelastic photoelectric effect (b) pair production and (c) resonant and non-resonant processes

Nuclear excitation by positron annihilation

A Feynman diagram for this process is shown in the lower part of *Fig. 9* (a) and (b). Diagram (a) represents the resonant process originally suggested by Present and Chen, and diagram (b) presents the more probable non-resonant process suggested by Raghavan and Mills [81Ra]. A resonant condition $| T + 2mc^2 - (\Delta E + B_K) | > \delta E/2$, has to be satisfied, T being the positron energy, mc^2 the electron rest mass, ΔE the energy, δE the width of the nuclear level and

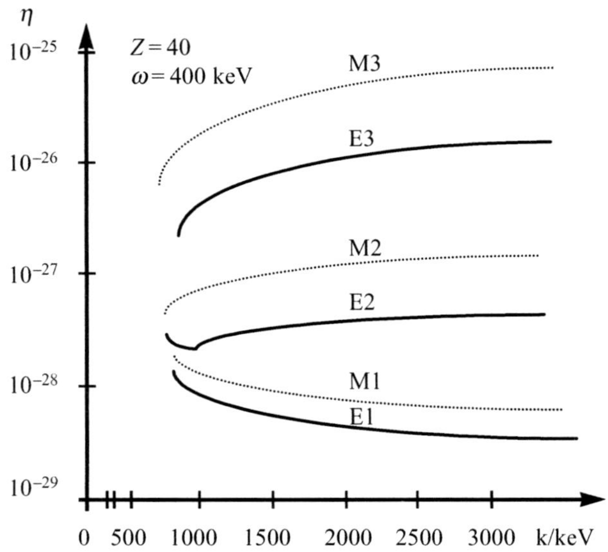

Fig. 10. The dependence of the cross-section for IPE on the transition energy for a given photon energy. $\eta = \sigma/g\Gamma_0$ (cm^2/keV)

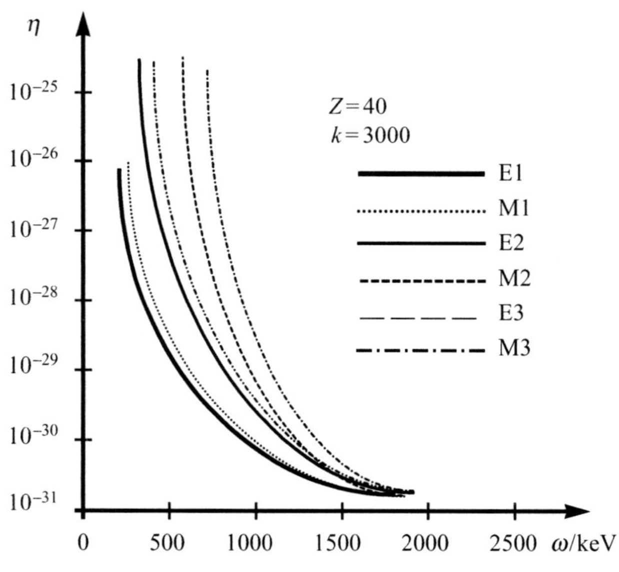

Fig. 11. The dependence of the total cross-section for IPE on the photon energy for a given transition energy. $\eta = \sigma/g\Gamma_0$ (cm^2/keV)

B_K is the K-shell-binding energy. Raghavan and Mills [81Ra] made the interesting suggestion that non-resonant contributions could resolve the discrepancy between theory and experimental practice.

Mukoyama and Shimizu [72Mu] in the first reported experimental investigation, irradiating ^{115}In with positrons of ^{22}Na source, observed conversion electrons from the 336 keV isomeric level; this is populated with a decay level of 1078 keV. They obtained a cross-section for this activation level which was seven orders of magnitude greater than the value predicted assuming an $E2$ transition.

They believed that in pair production, part (a) of *Fig. 9,* more positrons than estimated could contribute and the cross-section enormously enhanced. However Ljubicic et al. [84Lj] and Pisk et al. [85Pi] re-analysed this process and it is now clear that the cross-section for the non-resonant process is typically two orders of magnitude below the resonant value. The discrepancy between theory and experimental practice is still unresolved and this is a fertile field for further investigations.

The dependence of the total cross-section on IPE from a photon energy and a transition energy are shown in *Figs 10* and *11.* Total cross-section dependence on photon energy for E1 and M1 transitions is shown in *Fig. 12.*

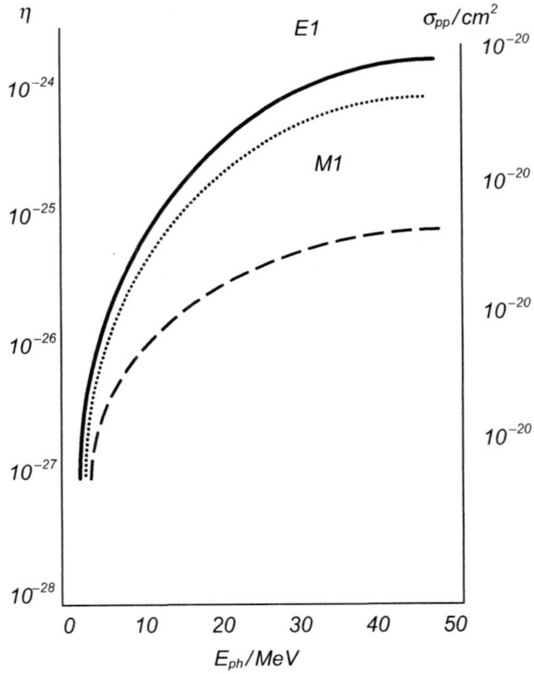

Fig. 12. Total cross-section dependence on the photon energy. Full curves represent the cross-section for E1, and M1 transition. The dashed curve represents the pair production cross-section, assuming $Z = 50$, $\eta = \sigma / g\Gamma_0$ (cm^2/keV)

An estimate was made in order to illustrate the magnitudes involved in the excitation of the giant dipole resonance, via this process, for ^{238}U at a photon energy of 15 MeV. Assuming a resonant energy of 10.96 MeV and a resonant width of 2.90 MeV [73Vey], Pisk and Ljubicic [87Pi] calculated a cross-section $\sigma_{E1} = 2\times10^{-30}$ cm^2; while it was measured to be 3×10^{-25} cm^2 for ^{238}U, at the same energy.

Nuclear de-excitation via electron-bridge mechanism

Contributions from higher-order transitions allow atomic electrons to influence γ-ray emission, with the effects being more pronounced for highly converted nuclear levels. The electron-bridge mechanism (EBM) can produce monoenergetic γ-rays with an energy E_γ, given by $E_\gamma = E_{tr} - (B_i - B_j)$, where E_{tr} is the transition energy, and B_i and B_j are the respective energies of the initial and final-state electrons. When $B_i > B_j$, photons with energies less than ω are emitted via an inelastic electron-bridge (IEB) effect. Kekez et al. [85Ke] made a theoretical estimate of the IEB effect with an evaluation of the Feynman diagram. The calculations involved the solving of numerically inhomogeneous Dirac equations for the nuclear Coulomb field. A hydrogen-like atomic model was assumed, although the Pauli exclusion principle was taken into account in the counting of the final electron states. The calculations were realistic and the external Coulomb field was included in all orders. Kekez et al. have studied the de-excitation of the 30.7 keV isomeric level of ^{93}Nb; they obtained an experimental value of 0.070 ± 0.018 for the relative probability of the inelastic electronic-bridge effect. This is within an order of magnitude of their theoretical prediction of 0.19.

The IEB effect produces a relatively small peak on a large background and therefore it requires a careful estimate of the background contributions. The contributions to the background from the internal Compton effect, and from the external *bremsstrahlung* associated with the internal conversion, have to be estimated with particular care.

4. Application of photoexcitation in astrophysical nucleosynthesis

It is now commonly accepted that the Universe as a whole constitutes one of the best laboratories dedicated to this scientific field. Moreover, nuclear and particle physics have taken, and are taking place throughout the history of the Universe.

Early nucleosynthesis occurred at a period during which the Universe had a temperature ranging from 10^9 K to 10^7 K. This is responsible for the formation of the lightest nuclear species (H, D, ^{3}H, ^{2}H, Li). For example, the ^{3}H(p, γ)^{4}He reaction produces γ quanta with an energy of 19.81 MeV. These and other γ-quanta, also produced in different nuclear reactions, may play an important role in the photoexcitation of the isomers of stable nuclei by nucleosynthetic processes.

Due to the fact that some of the influences of photoexcitation on the processes of the evolution of cosmos appear to be substantial, here a short survey is given of the main phases of the cosmo-chronology connected with the photoactivation of isomers.

4.1. Early nucleosynthesis

The "Big Bang" cosmological theory proposes that the whole Universe has been expanding for about 15 billions years and it was extremely hot and dense at the beginning of this period. This is supported by three independent facts. Hubble [29Hu] first observed the recession of the galaxies. Penzias and Wilson [65Pe] detected microwave background radiation of the universe. In the late 1940s, Alpher, Bethe and Gamow [48Al] considered the nucleosynthesis of the lightest chemical species (H, D, ^{4}He and ^{7}Li).

The Universe is also assumed to act as a perfect fluid specified by the scale function $R(t)$. These transformations are described by the Friedmann equation:

$$H^2 = \left(\frac{R}{R}\right)^2 = \frac{1}{3} 8\pi G\rho - \frac{k}{R^2} + \frac{1}{3}\Lambda, \qquad (22)$$

where Λ is the cosmological constant. If it is assumed that $\Lambda = 0$ and the so-called critical density ρ_c corresponding to $k = 0$ is given by:

$$\rho_c = \frac{3H^2}{8\pi G} = 1.88 \times 10^{-29}\ h_0^2\ \text{g cm}^{-3}, \tag{23}$$

where $h_0 = H_0/(100\ \text{km/s/Mpc})$, H_0 is equal to the present Hubble constant:

$$\Omega_N = \frac{\rho_n}{\rho_c} = 3.6 \times 10^{-3}\ h_0^{-2}\ \eta_{10} \left(\frac{T}{2.73}\right)^3, \tag{24}$$

where T is the present temperature of the Universe (in K) given by the thermal distribution of the microwave background radiation and $\eta_{10} = 10^{10}\ n_\beta/n_\gamma$, where n_β and n_γ are the baryon and photon densities. If $\Omega < 1$ the Universe is open, while it is closed if $\Omega > 1$.

The various periods of the Universe are illustrated schematically in *Fig. 13.* The upper abscissa shows the commencement and termination time of the periods and the lower arrow indicates the cooling of the successive phases.

Gamma ray lines are known to be produced in astrophysical objects such as supernovae, novae, neutron stars, black holes and interstellar media, when plasma temperatures exceed $10^8\ K$. Unique astrophysical information is encoded in the γ-ray lines – e.g. intensities, widths, and profiles contain information concerning abundance, bulk velocities, gravitational potential, densities, temperatures, the spectra of energetic particles, and magnetic field geometry.

Most of the earlier astronomical line observations were made in the respective optical and radio bands. However, from the onset of so-called "New

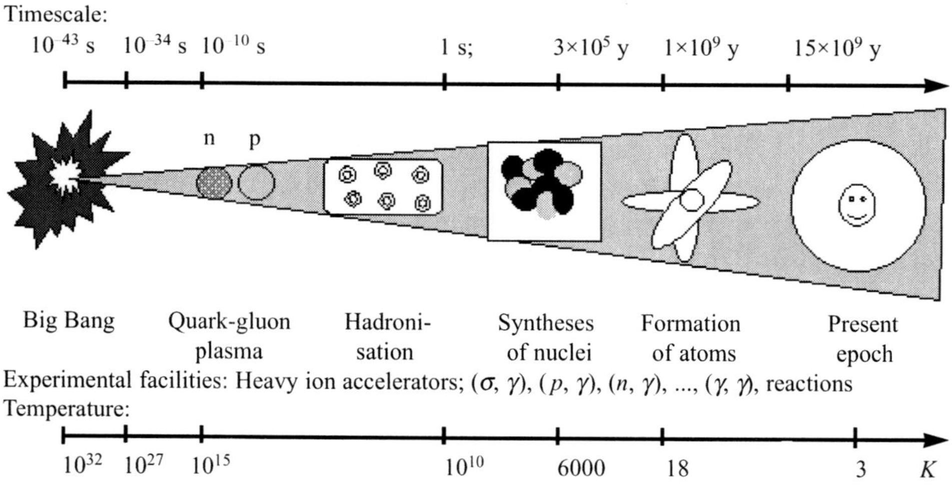

Fig. 13. Development path of the Universe from the Big Bang to the present epoch. The time and temperature scales are not proportional representations

Astronomies" unique information was provided by space infrared, ultraviolet, X- and γ-ray lines. At present, gamma-ray astronomy is experiencing a golden age. In the last ten years new γ-ray observatories have been constructed for measuring γ-ray intensities and creating an all-sky map up to 1.8 MeV by the COMTEL instrument [97Sc]. Gamma-ray lines are known to be produced in astrophysical objects by electron/positron annihilation, radioactive decay, nuclear excitation, neutron capture, and cyclotron processes in strong magnetic fields. The most important astronomical objects for γ-ray line observations are supernovae, novae, neutron stars, black holes and the interstellar medium. However, the sun can also be a γ-ray line source during strong solar flares. In view of the above, the possibility might seem to be given for reinvestigating the (γ, γ')-reaction experiment in accordance with the requirements of astrophysical studies.

By building the simplest Big Bang nucleosynthesis models, the following assumptions can be adopted. The Universe is assumed to be homogeneous and isotropic; its overall expansion or contraction takes place according to the General Relativity framework. As a consequence, its geometry is described by the Robertson–Walker metric:

$$ds^2 = dt^2 - R^2(t)\left\{\frac{d^2}{1-kr^2} + r^2\left(d\theta^2 + \sin\theta d\phi^2\right)\right\}. \tag{25}$$

The Standard Big Bang (SBB) model explains fairly well the observed abundance of the light elements up to mass number 7, where the nuclear processes involve only stable nuclei with relatively long lifetimes. Inhomogeneous Big Bang (IBB) models predict a considerable production of heavier elements. Most of the reaction cross-sections relevant for interpreting early universe nucleosynthesis in the framework of the SBB model are known within 10% uncertainty.

The elementary scheme of nucleosynthesis of the light elements assigns the sources of the light elements to those energy-generating stages of thermonuclear fusion in stars which terminate with a significantly increased abundance of the species in question. An abbreviated summary of the main line was elaborated by Clayton [68Cl] as follows:

- ^{4}He from hydrogen burning; ^{3}He from incomplete pp chain.
- D, Li, Be, B are bypassed, and their low abundances relative to the main line are interpreted as being due to non-thermal processes.
- ^{12}C and ^{16}O from helium burning ; ^{18}O and ^{22}Ne partially due to alpha capture by ^{14}N present during helium-burning processes.
- ^{14}N from conversion of CNO to ^{14}N in hydrogen burning, some ^{13}C, ^{15}N, and ^{17}O are also the result of CNO catalysis.

– ^{20}Ne, Na, Mg, Al, ^{28}Si – partly due to carbon burning.
– Mg, Al, Si, P, S – partly due to oxygen burning.

One can, with a fair degree of success, understand the relative abundance of the light elements in terms of the ashes of these burning stages. The success is hardly an unqualified one, however. Many abundance ratios make little sense in terms of a simple structure. A quantitative theory of the evolution of the galactic abundance of the light elements is badly obscured by fundamental uncertainties in the structural evolution of galaxies and in the details of mass loss in highly evolved stars. However, there remains the fundamental problem of why it is that the hydrogen-burning mass expelled from stars so greatly exceeds the helium-burning mass that is expelled. It seems that most of the nucleosynthesis is seen to occur in shells surrounding the cores of highly evolved stars; nevertheless, it is not at all clear how much may be expelled from the star at its death. Previous experimental work focused on the reactions of pp chains and the CNO cycles. Another frontier in nuclear astrophysics deals with the various astrophysical events in hot and explosive scenarios and the concomitant nucleosynthesis involving radioactive nuclei. The trend in this area can also be seen from the "Nuclei in the Cosmos" conferences (Crete, Greece, 1987; Baden/Vienna, Austria, 1990; Karlsruhe, Germany, 1992; Gran Sasso, Italy, 1994; Notre Dame, USA, 1996). In these, the percentage of contributions of experimental and theoretical work concerning nuclear processes involving unstable nuclei and/or astrophysical investigations in hot and explosive scenarios reached a level of almost 50 per cent. The cross-sections for proton-, and neutron-, alpha-capture reactions, and gamma-excitation on radioactive nuclei are extremely difficult to study [97Wi]. There have been only a limited number of such experiments being performed successfully.

It is likely that the conventional concept of stellar nucleosynthesis is not complete. Instead, the large abundance of ^{4}He may reflect the residue of a primordial fireball in the early Universe. On the other hand, perhaps one or more supermassive stars (or quasistars) have exploded at our galactic centre. (The reader may profitably turn to [65Fr], [66Pe], [67Cl], [68Cl], [82Ba], [82Cl], [98Al], [99Ha].) It seems certain that the thermonuclear burning stages are still relevant even if conventional stars turn out to be only a secondary source of nucleosynthesis.

The nuclear principles involved in the synthesis of heavier elements differ in many respects from the light ones. To illustrate the peculiarities of nuclear binding, a graph of nuclear energy per nucleon is displayed in *Fig. 14* as a function of the atomic weight. Quantitative examination shows that the true sequence of events follows a quite different process from the earlier burning

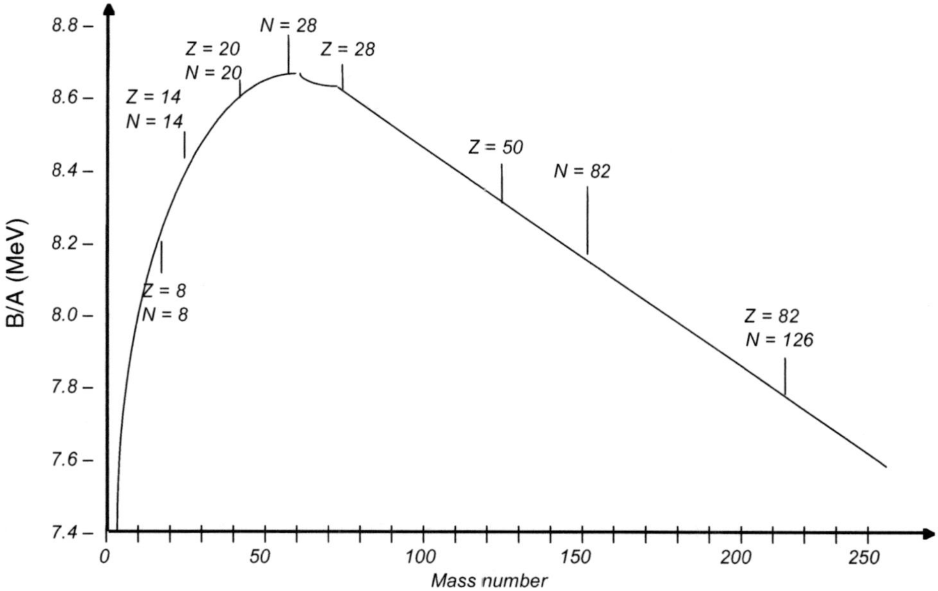

Fig. 14. The binding energy per nucleon of the most stable isobar of atomic weight *A* (After M. A Preston: *Physics of the Nucleus,* Addison-Wesley Publishing Company, Inc., Reading, Mass., 1962)

phases, and that the material again gravitationally contracts until nuclei begin to fuse with each other. The Coulomb barrier between two nuclei is so great that very high thermal temperatures would be required to effect their fusion. Before such high temperatures can be achieved in the contraction, the intense thermal γ-ray flux begins the photodisintegration of key nuclei, mainly by (γ, n), (γ, p), (γ, α), and (γ, γ') reactions. The photodisintegration rate is dominated by the - factor $\lambda_\gamma \propto \exp-(Q/kT)$, where Q is the binding energy of the particle (p, n, or α). The photoexcitation, (γ, γ') can produce an isomer state which can de-excite to the ground state of the same nucleus or decay to the adjacent nuclei by beta disintegration or electron capture. Consider what happens to the least tightly bound particles after their photoejection. They are quickly recaptured, perhaps to be photoejected once more. The inverse reactions attempt to establish an equilibrium, such that the number of photoejections per second is balanced by the number of recaptures per second. When the temperature rises, it may be described as a redistribution of loosely bound nucleons into more tightly bound states (this process is also referred to as the *photodisintegration rearrangement*).

The most tightly bound nuclei per nucleon are those with $50 < A < 60$, the maximum occurring at the ^{56}Fe nucleus. It may be noted here, however, that the binding per nucleon decreases again as yet heavier nuclei are approached.

The theoretically determined reaction rates are also often necessary to extrapolate experimental cross-sections from high energy measurements, down to the Gamow range of stellar energies. However, for the description of nuclear processes which are far from being stable (and where experimental data are scarcely or not at all available), at present one has to rely entirely on theoretical models.

The reaction mechanism and, therefore, also the respective reaction model depend on the level density in the compound nucleus (CN). The CN formalism describes two-step reaction proceedings over CN resonance, whereas the direct reaction formalism describes one-step reactions which do not involve CN-resonance. Considering cases where only a few isolated cases of resonance contribute the phenomenological *R*- or *K*-matrix, methods have often been employed to determine nuclear reaction rates, especially for reactions with compound nuclei in the mass $A = 8 - 28$ range. The cross-sections at thermonuclear energies are often determined by interfering non-resonant and resonant structures. In cases of high level density in the CN reaction, the cross-section is characterised by many instances of overlapping resonance, the CN mechanism dominates and the statistical Hauser–Fleshbach formalism can be applied.

For calculations of reaction cross-sections and reaction rates involving light nuclei, microscopic methods and few-body methods have been utilised. However, the fully microscopic or few-body approach is not reliable in producing phenomena which are physically observable and to which calculations of reaction rates are sensitive (such as particle thresholds, resonance and bound-state energies or scattering data). Nevertheless, microscopic and few-body methods have also been used successfully for investigating and extrapolating astrophysically relevant cross-sections.

The Distorted Wave-Born Approximation (DWBA) and Direct Capture (DC) models are also often used to obtain from indirect measurements the single-particle strength of resonance and bound states. This can then be used as an input for the calculation of reaction rates. Over the last few years, there has been enormous progress in nuclear structure models: Shell Model, Quasi-Particle Random Phase Approximation, Relativistic Mean Field Theory, Hartree–Fock–Bogoliubov Theory, and Finite Fermi System Theory.

These models are especially important for determining decay and reaction rates of radioactive nuclei where only very limited experimental information is available. Input parameters from nuclear-structure models for the calculation of nuclear decay and reaction rates include masses, level densities, deformations, excitation energies, spin/parity assignments, spectroscopic factors, density distributions, resonance energies and widths. One of the great challenges of the future is to check the reliability and to analyse the predictive power of the different nuclear structure models for determining these parameters for nuclei from the region of stability to the neutron and drip lines.

4.2. Processes in nucleosynthesis

In the pioneering work which was largely responsible for initiating the field of nuclear astrophysics, Burbidge et al. [57Bu] showed that the heavy elements are a consequence of three (*s*, *r* and *p*, respectively) essentially different processes:

– The *s-process*. When a nucleus is placed in a neutron flux, the nucleus captures a neutron, and it becomes an isotope of the same element with one unit more of atomic weight. ${}_ZX_N + n \rightarrow {}_ZX_{N+1}$ $(Z, A \rightarrow Z, A+1)$, where $Z+N = A$. If the nucleus $(Z, A+1)$ is stable, it waits until it captures another neutron, and so on. However, if the nucleus $(Z, A+1)$ is radioactive, the question as to whether it β^--decays to $(Z+1, A+1)$ or captures a second neutron depends upon the relative lifetimes. If the β^--decay is faster than the second capture of the neutron, then this type of neutron capture is called the *s*-process (slow-process). The *s*-process path is frequently said to pass along the valley of beta-stability, which means that it sticks closely to the most tightly bound isomer of each atomic weight.

– The *r-process*. The classical *r*-process requires the neutron capture rates to be much faster than those for β-decay. Some astrophysical environments have produced very neutron-rich nuclei, and after the synthesising event these nuclei undergo chains of β^--decay; these terminate at the most neutron-rich of the stable isobars, at each value of atomic weight. This decay chain is indicated by the letter *r*.

– The *p-process*. Some proton-rich nuclei cannot be synthesised by either the slow-neutron-capture chain or by the decay of neutron-rich matter. These nuclei are shielded from both processes and are labelled with the letter *p* in deference to their proton-rich character.

– The *e-process*. This process plays an important role in the dominance of the ratio of ^{54}Fe to ^{56}Ni in a gas in complete equilibrium at one temperature and density. The *e*-process in stars has either occurred so rapidly that the dominant nucleus produced is ^{56}Ni (which decays later to ^{56}Fe) or has been produced so slowly that beta decays have reduced *Z/N* to a value so low that ^{56}Fe is the dominant equilibrium nucleus.

– The *ν-process*. As the core of a massive star collapses to form a neutron star, the flux of neutrons becomes so great that, despite the small cross-section, substantial nuclear transmutation is produced by ν-nucleosynthesis.

In several papers – [74Ed], [74Ha], [75Ha] – the question has been raised that the low-temperature photonuclear nucleosynthesis may play a part in the production of several rare nuclei, the origins of which do not seem to be adequately accounted for in most traditional *s*-, *r*-, *p*- and some other processes.

To get a better understanding of how this distinction is made operationally, it will be helpful to examine *Fig. 15,* which is a schematic part of a nuclide chart.

The proton capture reaction for low-temperature photonuclear processing is the $^{3}T(p, \gamma)^{4}$He reaction, where the energy of γ-ray is $19.8142 + E_p$ MeV. This reaction would not be expected to operate for any length of time during the

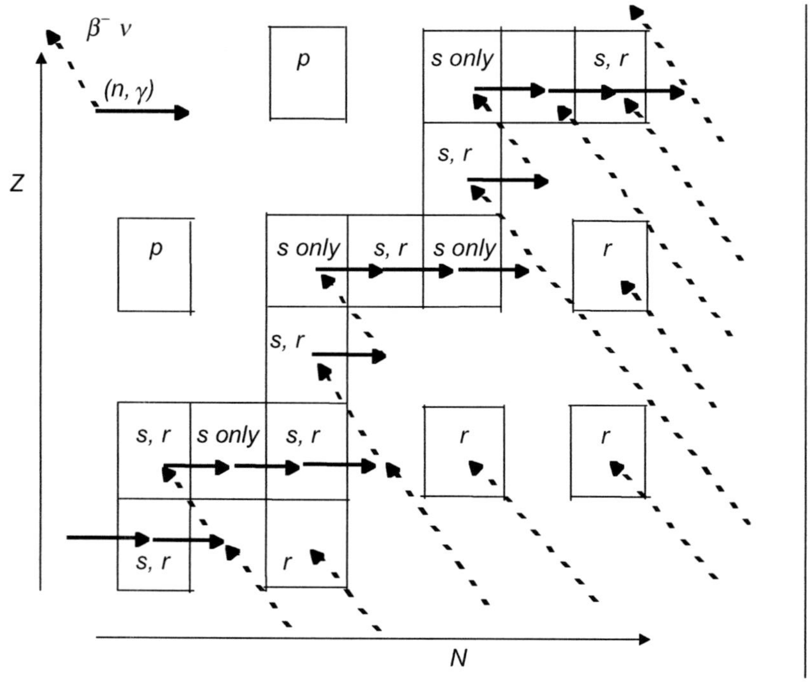

Fig. 15. A characterisation of a portion of the chart of nuclides showing the assignment of nuclei to classes of the *s*-, *r*-, and *p*-processes, respectively

ageing of the star from its main-sequence lifetime to its red giant stage, since tritium burning first requires electron capture on He-3 via ^{3}He$(e^{-})^{3}T$; this can only occur in a degenerate hydrogen-burning region. A branching ratio for the competition between electron capture by ^{3}He the ^{3}He(^{3}He, $2p$) ^{4}He and, at higher temperatures, the ^{3}He(^{4}He, γ)^{7}Be is given by:

$$\frac{^{3}\mathrm{He}(e^{-})^{3}\mathrm{T}(p, \gamma)^{4}\mathrm{He}}{^{3}\mathrm{He}(^{3}\mathrm{He}, 2p)^{4}\mathrm{He} + {}^{3}\mathrm{He}(^{4}\mathrm{He}, \gamma)^{7}\mathrm{Be}}$$

or, more conveniently in terms of the mean lifetimes of each (designated by the τ_{3e} for electron capture and τ_{33} and τ_{34} for the normal ^{3}He capture path).

High-energy photons will either Compton-scatter – via nuclear capture – or trigger photonuclear events. The importance of the Compton-scattering cross-section must be emphasised, in that it effectively determines the relationship between the exposure and the hydrogen-burning interval. The long-lived isomeric states also play an important role in the *s*-process, because these metastable states are often significantly populated by either the γ-cascades following neutron capture on the preceding isotope, or directly by the β-decay of an unstable isobar. The criteria for consideration will be that the state (i) undergoes quick β-decay at a rate comparable to its spontaneous radiative-decay rate to the ground state, or (ii) has such a long life compared to the lifetime of the ground state that it could capture a neutron before decaying.

The minimum values are placed in the temperature and density range from $\rho = 8000 - 22000$ g cm^{-3}, with the corresponding limit to the temperature range being $T = (5-20) \times 10^6$ K. These high-energy photons will, either via electron Compton-scatter (see equation 9) or via nuclear capture, trigger photonuclear events by a (γ, n) or (γ, $2n$) reaction. The cross-sections of the Compton-scatter and the (γ, n) and (γ, $2n$) are 25 mb and average 50–70 mb, respectively. *Fig 16* shows a comparison between the magnitude of the average cross-sections of (γ, n) and (γ, $2n$) as a function of target nucleus mass, evaluated at 20 MeV.

A typical example of (γ, n) and (γ, $2n$) resonant shapes and their relationship to 20 MeV photon scattering probability curve (ECS) – as determined by electron Compton-scatter – is shown in *Fig. 17*. The cross-section curves are usually in millibarns, whereas, to the left, for scattering probability they are in $k_0^2\phi_R dk / (\pi r_0^2 \mu dk)$ units where k_0 is the initial photon energy (MeV), ϕ_R is the

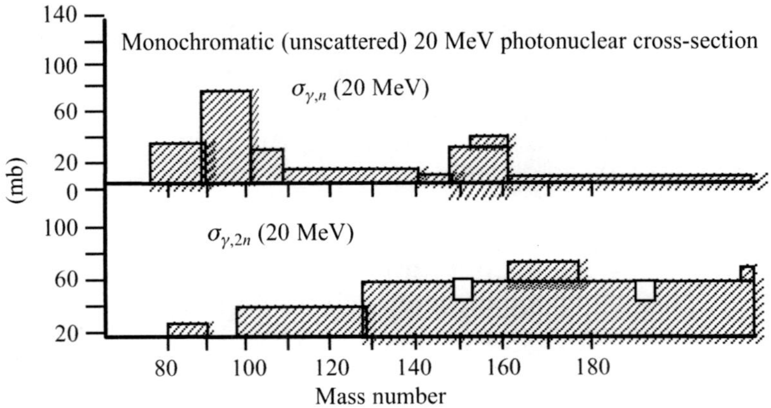

Fig. 16. A comparison between the magnitude of the (γ, n) and (γ, $2n$) average cross-section as a function of target nucleus mass, evaluated at 20 MeV [78Ha]

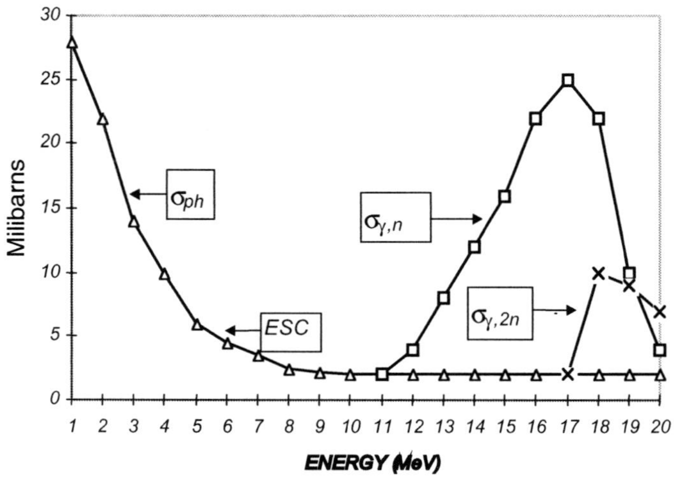

Fig. 17. An example of (γ, n), $(\gamma, 2n)$ resonant shapes and their relationship to the 20 MeV photon-scattering probability curve (ESC), as determined by electron Compton-scatter

relative photon flux at energy k, r_0 is the classical electronic radius (in cms), and μ is the electron mass (mc^2). These curves are only intended to be typical, and do not necessarily depict a specific case [78Ha]. Harrison suggested that with the occurrence of degenerate hydrogen burning in white dwarfs accreting hydrogen-rich material, and the subsequent closure of p–p cycle via $^3He(e^-)^3T(p, \gamma)^4He$, the energetic photon ($E_\gamma = 20$ MeV) thus emitted may trigger (γ, n) and $(\gamma, 2n)$ reactions by *s*- and *r*-seed nuclei; this leads to the synthesis of the traditional nuclei.

4.3. Nucleosynthetic effects of γ-fluxes at stellar surfaces and in the solar system

It is useful to examine the possibilities of the applicability of isomer excitation by γ-rays produced in the galaxies.

4.3.1. Production of 1 10 MeV *γ-rays* by high energy particles

High energy protons produce γ-rays of energy greater than 1 MeV in stellar flares and also as a by-product of the galactic cosmic rays. Three kinds of events used to be distinguished:

a) "Instantaneous" photons produced at the same time as the flare or cosmic ray irradiation. These photons are mainly produced by inelastic reactions with the Si and Fe group. In the stellar flares the γ-rays of 2.23 MeV (produced by the thermalised neutrons absorbed by hydrogen) represent a large part of the instantaneous photons.
b) "Delayed" photons produced in the decay of some unstable nuclides such as ^{10}C, ^{7}Be, ^{22}Na or ^{24}Na.
c) Positrons e^+ produced by some β^+ emitters, such as ^{11}C, ^{13}N, ^{15}O (whose decay results in 0.511 MeV γ-rays).

Table 6. Production of γ-rays in stellar flares.
10, 20, 50 and 150 MeV γ-rays are produced by protons

$E\gamma$ (MeV)	Target	E_p = 10 MeV		E_p = 20 MeV		E_p = 50 MeV		E_p = 150 MeV	
		σ (mb)	σ^* (mb)	σ (mb)	σ^* (mb)	σ (mb)	σ^* (mb)	σ (mb)	σ^* (mb)
<1.25	Fe	1000	3.4×10^{-2}	1000	3.4×10^{-2}	700	2.4×10^{-2}	300	1×10^{-2}
≤2	Mg–Si	500	5×10^{-2}	300	3×10^{-2}	200	2×10^{-2}	50	5×10^{-3}
1.6	^{14}N	60		10		15		10	
1.6	^{16}O	0	2.2×10^{-2}	7	1.6×10^{-2}	10	1.4×10^{-2}	5	7.2×10^{-2}
1.6	^{20}Ne	200		100		40		20	
2	^{12}C		0	10	5×10^{-3}	30	1.5×10^{-2}	6	3×10^{-3}
2.3	^{4}He					60		60	
2.3	^{14}N	4	3.8×10^{-4}	10	1.1×10^{-2}	15	2	10	2
2.3	^{16}O			10		10		10	
2.3	^{20}Ne			5		10		5	
4.4	^{12}C	350		100		30		15	
4.4	^{14}N		0.175	10	8.2×10^{-2}	30	6.9×10^{-2}	20	2.4×10^{-2}
4.4	^{16}O			30		55		15	
4.4	^{20}Ne			5		10		5	
5.2	^{16}O	0	0	40		50		10	
5.2	^{20}Ne			20	3.8×10^{-2}	10	4.6×10^{-2}	10	9.8×10^{-2}
6.1	^{16}O	300	0.273	70	6.9×10^{-2}	80	7.4×10^{-2}	30	2.8×10^{-2}
6.1	^{20}Ne			60		20		10	
6.6	^{12}C	0	0	30	1.5×10^{-2}	10	5×10^{-2}	10	5×10^{-2}
7.1	^{16}O	100	9.1×10^{-2}	40	3.8×10^{-2}	20	1.9×10^{-2}	10	9.8×10^{-2}
7.1	^{20}Ne			20		10		10	

The target which is responsible for the formation is indicated. The reaction σ (mb) and the equivalent cross section σ^* (mb) = $\sigma\, n_i/n_H$ are given for each considered photon energy (from [70Au])

By a simple use of formation cross-sections for γ-rays, the spectrum of the instantaneous photon fluxes would be given when they are produced in a flare, or by the galactic cosmic rays. In the special case of the flares, where the number of high energy protons released is known, the photon and positron fluxes emitted by these events can be calculated. The main γ-rays and their respective origins are given in *Table 6.*

Audouze [70Au] computed the rate of isotope formation generated when a flux of fast proton irradiates material of various compositions. For the calculation he used the work of Lingenfelter and Ramaty [67Li], together with the report of Zobel et al. [67Zo]. It was found that in the majority of solar flares the calculated photon fluxes are larger than the observed limit.

The shape of integral photon fluxes emitted in a flare is:

$$\frac{\phi_\gamma\left(E_\gamma \geq E_0\right)}{\left(E_\gamma \geq E'\right)} = \frac{\Sigma_i \sigma_i^*\left(E_\gamma \geq E_0\right)}{\Sigma_i \sigma_i\left(E_\gamma \geq E'\right)}, \tag{26}$$

where $\sigma_i^* = \sigma\, n_i/n_H$ is the effective cross-section of γ-ray production from the target i. E_0 and E' are the γ-ray energies varying from 1 to 6 MeV, respectively, and the reference energy is $E'_\gamma = 4$ MeV.

In the low flux approximation ($\sigma_d \phi T << 1$) the $n_i/n_H = \Sigma_i \sigma_i \phi(Q_{ij})t$. In other words, if the proton flux is not intense enough to destroy j or i, then the spallative abundance is simply proportional to the integral flux (as shown in region a) of *Fig. 18*).

In a high flux approximation two destruction terms become important while others remain negligible: for instance, $\sigma_{j\,\text{des}}\phi(Q_j) >> 1$, while $\sigma_{i\,\text{des}}\phi(Q_i)T << 1$. In these condition the spallative abundance n_j tends to the asymptotical value:

$$\frac{n_j}{n_H} \to \frac{\sum_i \frac{n_i}{n_H} \sigma_{ij}\, \phi\left(Q_{ij}\right)}{\sigma_{i\,\text{des}}\, \phi\left(Q_j\right)}. \tag{27}$$

This is illustrated on *Fig. 18* by region b).

Finally, for very intense fluxes the abundance of the targets and the products decrease together (region c) of *Fig. 18*).

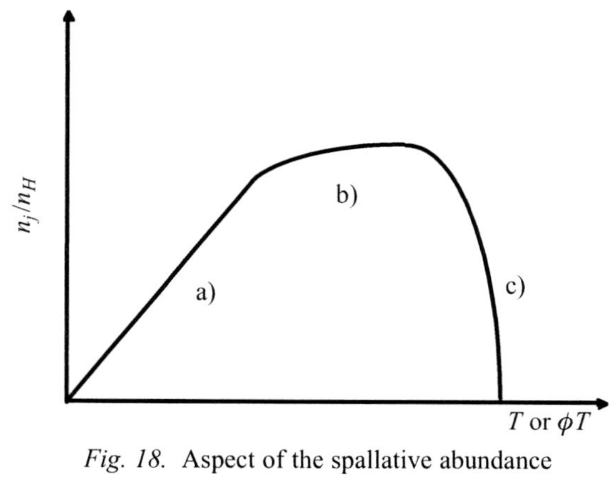

Fig. 18. Aspect of the spallative abundance of the j element with respect to the integrated flux $\phi(Q)t$

The production of γ-rays by galactic cosmic rays (with energies, targets, and cross-sections σ and σ^*) are shown in *Table 7.* The reactions induced by α-particles as well as protons are considered. All these incident particles are supposed in concert with E_p or the E_α>1 GeV/nucleon.

To get rid of the difficulties caused by poor knowledge of flare geometry, a formula was given in Ryter et al. [70Ry]. It is used to relate the number of photons n_γ or positrons n_{e+} produced, with the corpuscular energy E_T released in the flare:

$$n_{\gamma,e+} = \frac{E_T \int_{Q_e}^{\infty} \sigma_{e\gamma} \frac{n_e}{n_H} \phi \, dE}{\int_0^{\infty} \varepsilon \phi \, dE}, \tag{28}$$

where Q_c and $\sigma_{e\gamma}$ are the threshold and the cross-section of the production reaction and ε the stopping power of the protons. The $\varepsilon = A/E_p$ (with ε in MeV/particle cm^{-3}, $A = 2\times10^{-21}$ if E_0 is expressed in MeV.

Table 7. Production of γ-rays by galactic cosmic rays

E_γ (MeV)	Target	σ_p (mb)	σ_{He} (mb)	Diff. flux [a)]
<1	Fe peak	150	400	0.1
1–2	CNO	10		
1–2	Mg-S	50	200	
1–2	Fe peak	150	400	1
2–3	CNO	10		
2–3	Mg-S	5	23	0.3
3–4	CNO	5	10	
3–4	Mg-S	10	60	0.3
4–5	CNO	20	40	0.6
5–6	CNO	10	15	0.3
6–7	CNO	20	20	0.6
>7	CNO	10	10	0.3

a) arbitrary units

The effect of galactic cosmic rays also needs to be considered in the measurements. Vette et al. [69Vet] show that the γ-ray spectrum is flat for E_γ between 2 and 6 MeV. The spectrum decreases very slowly between 1–7 MeV and may be considered as rather flat. The formation of γ between 1 and 3 MeV is due to the large contribution of the Fe and Si group. The theoretical photon-spectrum shape looks like the observed flux but does not have sufficient intensity. Fowler et al. [70Fo] have calculated a low contribution for the observed

γ-ray flux. In *Table 7* the γ-ray energies, targets, σ and σ^* cross-sections are given to the reactions induced by $\alpha - s$ and protons. All these incident particles are supposed to be in concert with E_p or the $E_\alpha > 1$ GeV/nucleon.

4.3.2. The gamma-ray sky instruments

The Compton Telescope (COMPTEL) unit has four experimental instruments: (i) The Burst And Transient Source Experiment (BATSE) is optimised to measure brightness variation on time-scales down to several microseconds, over the energy range 30 keV to 1.9 MeV; (ii) The Oriented Scintillation Spectrometer Experiment (OSSE) is designed to undertake comprehensive observation of astrophysical sources in the 0.1 to 10 MeV range; (iii) The Energetic Gamma-Ray Telescope (EGRET) is the highest energy instrument on the Compton telescope, and covers the broadest energy range: from 10 MeV to 30 GeV; (iv) The Imaging Compton Telescope detects gamma-rays by registering the occurrence of two successive interactions in the telescope. The OSSE and BATSE are American instruments; COMPTEL and EGRET were built within the framework of large international collaboration. The institutes involved are listed in *Table 8.*

Table 8. Four Gamma-Ray Observatories (GRO)

Abbreviation	Telescope	Laboratory
OSSE	Oriented Scintillation Spectrometer Experiment	Naval Research Laboratory (USA); North Western University
COMPTEL	Compton Telescope	Max-Planck-Institute für extraterr. Physik (FRG); Space Research Organisation of the Netherlands; University of New Hampshire (USA); Space Science Department of ESA (NL)
EGRET	Energetic Gamma-Ray Telescope Experiment	Goddard Space Flight Center (USA); Max-Planck-Institute für extraterr. Physik (FRG); Stanford University (USA)
BATSE	Burst And Transient Source Experiment	Marshall Space Flight Center (USA); Goddard Space Flight Center (USA); University of California, San Diego (USA)

The first EGRET catalogue [94Fi] covered the first cycle of the Compton Gamma-Ray Observatory (CGRO); this began on 5 April 1991 and ended in the same year on 17 November. The catalogue included one solar flare, pulsars, gamma-ray bursts, a normal galaxy, an active galaxy and undefined sources. Compton-scattering collision occurs in a detector of low $-Z$ material in one of the $D1$ modules, then a second interaction takes place in a detector with high $-Z$ material in the $D2$ module; ideally, in this the scattered gamma-ray is totally absorbed [93Sc], [97Sc].

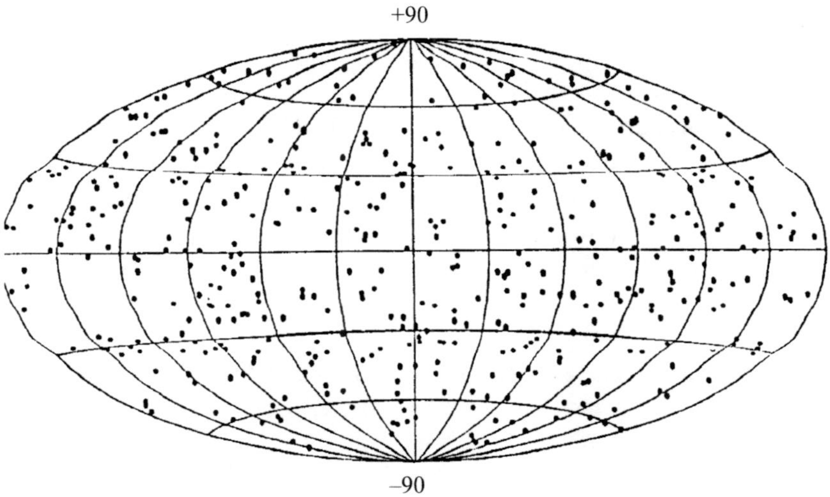

Fig. 19. Location of the first 447 γ-ray burst seen by the BATSE instrument, plotted using galactic coordinates (from [94Kn]

The telescope covers the energy range from 30 MeV to over 20 GeV. EGRET records γ-rays photons individually as electron-positron pair-production events. These are processed automatically to provide the arrival direction and energy of each photon. The point-spread function (PSF) is energy dependent, having a FWHM of approximately 6° at 100 MeV and smaller values at higher energies. Hartman et al. [99Ha] published the third catalogue of high-energy γ-ray sources detected by the EGRET telescope at the Compton Gamma-Ray Observatory (CGRO).

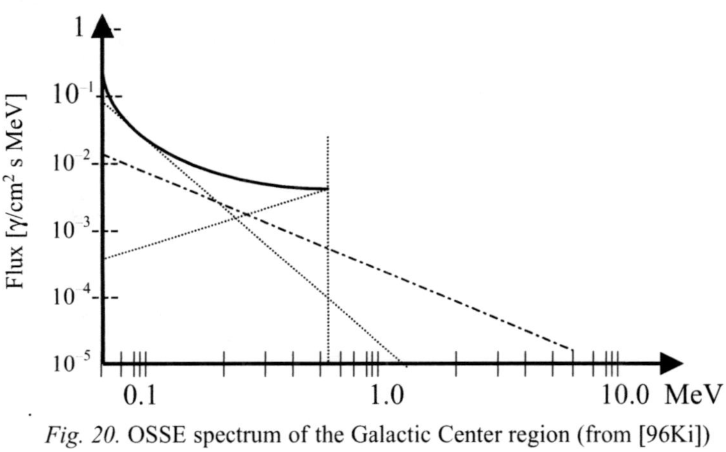

Fig. 20. OSSE spectrum of the Galactic Center region (from [96Ki])

A total of 588 cosmic gamma-ray bursts have been detected by the BATSE [94Kn]. Meegan et al. [92Me] have also found γ-ray bursts *(Fig. 19).*

The annihilation (511 keV) line has been measured by OSSE with good statistics from the general direction of the galactic center region. The energy spectrum between 50 keV and about 4 MeV clearly shows three components of a low-power continuum emission, the 511 keV line feature, and three-photon continuum emission from positronium formation (see *Fig. 20*). The best estimate of the positronium fraction is 0.97±0.03.

4.4. Thermalisation of long-lived isomeric states under stellar conditions

During the astrophysical studies an important task was to examine the various paths and mechanisms by which the isomeric states of nuclei are formed and brought into statistical thermal equilibrium with their ground state. Later (section 6), a quantitative examination will be made of the internal nuclear equilibration both by direct linking between an excited state and the ground state, and by direct linking via an intermediate state for the important case of long-lived isomeric states. To ascertain whether or not the ground state and the isomer are able to maintain their separate identities under stellar conditions, the formalism of Ward and Fowler [80Wa] has been used. First of all, the simple two- and three-level systems are looked at. The main standpoint of these considerations will be to concentrate on the internal nuclear details of many intermediate states of interest in order to obtain the relevant scales as a function of stellar temperature (required to establish a thermal distribution of excited-state populations).

4.4.1. Equilibration of an excited state by direct transitions

In *Fig. 21* an illustration is given of the various modes of production by direct transitions. The time evolution of the abundance of n_0 and n_e (cm^3) of the ground and excited states, respectively, is described by the coupled system:

$$\frac{dn_0}{dt} = \lambda_{p0} n_p - (\lambda_{0e} + \lambda_{0d}) n_0 + \lambda_{e0} n_e ,$$

$$\frac{dn_e}{dt} = \lambda_{pe} n_p + \lambda_{0e} n_0 - (\lambda_{0e} + \lambda_{ed}) n_e . \tag{29}$$

The easiest way to solve the above set of equations is to assume that the population n_p, and all the rates λ, are positive constants in time. If all the other populations are set initially to zero, then it is possible to obtain a time-independent solution:

$$n_0(t) = \frac{n_p}{s_1 s_2}\left[\lambda_{e0}\left(\lambda_{p0} + \lambda_{pe}\right) + \lambda_{ed}\lambda_{p0}\right] + \\ + \frac{n_p}{\left(s_2 - s_1\right)} \sum_{j=1}^{2} \frac{(-1)^j}{s_j} \exp\left(s_j t\right)\left[\lambda_{p0}\left(s_j + \lambda_{e0} + \lambda_{ed}\right) + \lambda_{pe}\lambda_{ed}\right], \tag{30}$$

where s_1 and s_2 are the two distinct negative roots of the quadratic polynomial:

$$Q(s) \equiv s^2 + \left(\lambda_{e0} + \lambda_{0e} + \lambda_{ed} + \lambda_{0d}\right)s + \lambda_{0d}\left(\lambda_{e0} + \lambda_{ed}\right) + \lambda_{0e}\lambda_{ed}. \tag{31}$$

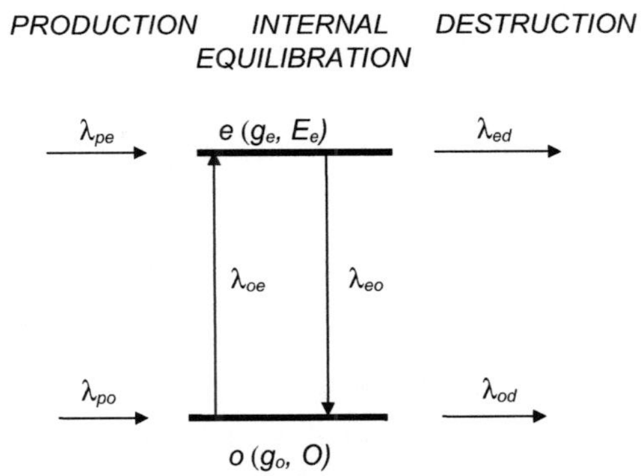

Fig. 21. A schematic picture of a single excitation/de-excitation mechanism of internal equilibration

Since equations (29) and (31) are unchanged with respect to interchanging the subscripts 0 and e, the analogous solution for $n_e(t)$ can be obtained from equation (30) by the substitutions $0 \leftrightarrow e$. From equation (30) and its equivalent for $n_e(t)$ it can be seen that the ratio $n_e(t) / n_0(t)$ is independent of the constant total source production rate, which equals $\lambda_{p0} + \lambda_{pe}$, but depends on the production branching ratios:

$$f_{p0} = 1 - f_{pe} \equiv \frac{\lambda_{p0}}{\lambda_{p0} + \lambda_{pe}}. \tag{32}$$

Taking the steady-state limit of equation (30) with $t = \infty$ gives the ratio:

$$\left(\frac{n_e}{n_0}\right)_{t\to\infty} = \frac{\lambda_{0e} + f_{pe}\lambda_{0d}}{\lambda_{e0} + f_{p0}}. \tag{33}$$

For rapid destruction ($f_{pe}\lambda_{0d} >> \lambda_{0e}$ and $f_{p0}\lambda_{ed} >> \lambda_{e0}$), the steady-state abundance ratio is determined by the source production ratios and the production rates as:

$$\left(\frac{n_e}{n_0}\right)_{t\to\infty} = \frac{f_{pe}\,\lambda_{0d}}{f_{p0}\,f_{ed}}. \tag{34}$$

For most nuclear excited states that are not isomers, it will generally be the case that $\lambda_{e0} >> \lambda_{ed}$, since the lifetimes of such excited states are typically in the nanosecond to the picosecond range. Under such intermediate circumstances:

$$\left(\frac{n_e}{n_0}\right)_{t\to\infty} = \left(\frac{\lambda_{0e}}{\lambda_{e0}}\right)\left(\frac{f_{pe}\lambda_{0d}}{\lambda_{e0}}\right) >> \left(\frac{\lambda_{0e}}{\lambda_{e0}}\right) \tag{35}$$

which exceeds the thermal population ratio:

$$\left(\frac{n_e}{n_0}\right)_{t\to\infty} = \frac{\lambda_{0e}}{\lambda_{e0}} = \frac{g_e}{g_0}\exp\left(\frac{E_e}{kT}\right). \tag{36}$$

For the nuclear destruction, $\lambda_{0d} = \lambda_{ed} \to 0$ in equations (29) and (31) causes $Q(s) \to s(s + \lambda_{e0} + \lambda_{0e})$, and a time-dependent solution that characterises how quickly internal equilibration operates is explicitly:

$$n_0(t) = \frac{n_p\lambda_{p0}\left(\lambda_{p0}+\lambda_{pe}\right)t}{\lambda_{e0}\lambda_0 e} + n_p\left(\frac{\lambda_{pe}\lambda_{e0} - \lambda_{p0}\lambda_{0e}}{(\lambda_{e0}+\lambda_{0e})^2}\right)\left\{\exp\left[-(\lambda_{0e}+\lambda_{e0})\,t\right]-1\right\}. \tag{37}$$

If stimulated emission due to the hot stellar photon bath is only enhancement to the excited state's spontaneous EM transition rate (which equals $\lambda_{e0}{}^{s}$), then the internal nuclear equilibration time is given by:

$$\tau_{eq} = \frac{1-\exp\left(-E_e / kT\right)}{1+\left(g_e / g_0\right)\exp\left(-E_e / kT\right)}\left(\frac{1}{\lambda^{s}_{e0}}\right) \Rightarrow \left(\frac{1}{\lambda^{s}_{e0}}\right) \qquad \text{for } kT \ll E_e,$$

$$\text{and} \quad \Rightarrow \left(\frac{E_e / kT}{\left(1+g_e / g_0\right)\lambda_{e0}\,s}\right) \qquad \text{for } kT \gg E_e, \tag{38}$$

where there is an explicit introduction of the factor which translates the spontaneous reaction rate into that, including the stimulated emission.

If the simplified two-level ground-state system is considered to be that "target" for the destruction rates λ_{0d} and λ_{ed}, the total steady-state destruction rate (λ_{ss}) may be defined as:

$$\lambda_{ss} = \frac{n_0(t\to\infty)\lambda_{0d} + n_e(t-\infty)\lambda_{ed}}{n_0(t-\infty)+n_e(t-\infty)} = \frac{\lambda_{0d}\left(\lambda_{e0}+f_{0e}\lambda_{ed}\right)+\lambda_{ed}\left(\lambda_{0e}+f_{pe}\lambda_{0d}\right)}{\lambda_{e0}+\lambda_{0e}+f_{po}\lambda_{ed}+f_{p0}\lambda_{0d}}\ . \tag{39}$$

For the case of rapid internal equilibration, equation (33) shows that equation (39) reduces to:

$$\lambda_{ss} \Rightarrow \frac{g_0 \lambda_{0d} + g_e \exp\left(-E_e / kT\right)}{g_0 + g_e \exp\left(-E_e / kT\right)}, \tag{40}$$

where the denominator is just the partition function for a two-level system. In the opposite, extreme system of rapid nuclear destruction, equation (34) implies that the steady-state destruction decay rate becomes:

$$\lambda_{ss} \Rightarrow \frac{\lambda_{ed} \lambda_{0d}}{f_{p0} \lambda_{ed} + f_{pe} \lambda_{0d}}. \tag{41}$$

From the last two equations it can be seen that if the nuclear destruction rates are sufficiently rapid, a substantial deviation from the usual thermal-mixture "stellar reaction rate" can occur. For the special case of long-lived isomers their spontaneous decay is not sufficiently rapid to allow direct internal coupling to the ground state.

4.4.2. Equilibration through an intermediate state

The three-level excitation and de-excitation scheme for indirectly linking long-lived isomers to their ground state is shown schematically in *Fig. 22.* It is assumed that no direct transition takes place between the isomer (m) and the ground state; thus they are only linked indirectly via intermediate level (i) with the transition

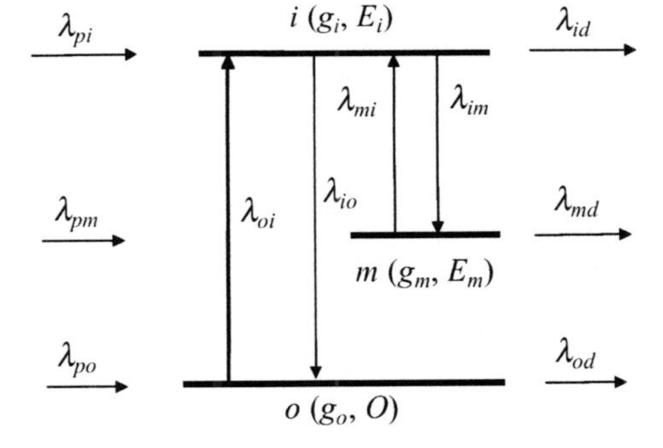

Fig. 22. A schematic picture of an internal excitation/de-excitation mechanism of the equilibration in a three-level nuclear system

rates shown. Each of the three levels is fed at a constant source production rate: λ_{pk}, $k = 0, m, i$. Each level can also decay at an indicated nuclear destruction rate: λ_{kd}. The coupled set of differential equations describing the time evolution of the abundance is:

$$\frac{dn_0}{dt} = \lambda_{p0}\, n_p - (\lambda_{01} + \lambda_{0d})\, n_0 + \lambda_{i0}\, n_i\,,$$

$$\frac{dn_m}{dt} = \lambda_{pm}\, n_p - (\lambda_{m1} + \lambda_{md})\, n_m + \lambda_{im}\, n_i\,, \tag{42}$$

$$\frac{dn_i}{dt} = \lambda_{pi}\, n_p - \lambda_{0i}\, n_0 + \lambda_{mi}\, n_m - (\lambda_{i0} + \lambda i_m + \lambda_{id})\, n_i\,.$$

If it is assumed that the situation is the same as in the previous case, the solution to equation (42) can be written as:

$$\begin{aligned} n_0(t) = \frac{-n_p}{s_1\, s_2\, s_3} \{ [\lambda_{i0} (\lambda_{p0} + \lambda_{pm} + \lambda_{pi}) + \lambda_{p0} \lambda_{id}] \lambda_{mi} + \\ + [\lambda_{p0} (\lambda_{i0} + \lambda_{im} + \lambda_{id}) + \lambda_{pi}\, \lambda_{i0}] \lambda_{md} \} + \\ + n_p \sum_{j=1}^{3} \frac{\exp(s_j t)}{s_j} \left[\prod_{l=1,\, l \neq j}^{3} \frac{1}{s_j - s_l} \right] \times \{ [\lambda_{i0} (\lambda_{p0} + \lambda_{pm} + \lambda_{pi}) + \lambda_{p0} (s_j + \lambda_{id})] \lambda_{mi} + \\ + [\lambda_{p0} (s_j + \lambda_{i0} + \lambda_{im} + \lambda_{id}) + \lambda_{pi} \lambda_{i0}] (s_j + \lambda_{md}) \}, \end{aligned} \tag{43}$$

where the three quantities s_1, s_2, s_3 (assumed to be distinct and less than zero) are roots of the cubic polynomial:

$$\begin{aligned} C(s) \equiv (s + \lambda_{i0} + \lambda_{im} + \lambda_{id})\,(s + \lambda_{0i} + \lambda_{0d})(s + \lambda_{mi} + \lambda_{md}) - \\ - \lambda_{im} \lambda_{mi}\, (s + \lambda_{0i} + \lambda_{0d}) - \lambda_{i0}\, \lambda_{0i}\, (s + \lambda_{mi} + \lambda_{md}). \end{aligned} \tag{44}$$

The analogous expression for $n_m(t)$ can be obtained by exchanging the subscripts 0 and m. The steady-state solution is:

$$\left(\frac{n_m}{n_0} \right)_{t-\infty} = \frac{(f_{im} + f_{pm}\, f_{id})\, \lambda_{0i} + (f_{pm} + f_{pi}\, f_{im})\, \lambda_{0d}}{(f_{i0} + f_{p0}\, f_{id})\, \lambda_{mi} + (f_{p0} + f_{pi}\, f_{i0})\, \lambda_{md}}\,, \tag{45}$$

where the additional fractional branching from the intermediate coupling level is defined as

$$f_{ix} \equiv \frac{\lambda_{py}}{\lambda_{i0} + \lambda_{im} + \lambda_{id}} \quad \text{for } x = 0, m, d, \qquad f_{py} \equiv \frac{\lambda_{py}}{\lambda_{p0} + \lambda_{pm} + \lambda_{pi}} \quad \text{for } y = 0, m, i. \tag{46}$$

If the nuclear destruction rates are dominant over all other internal equilibration rates, then the equation of (45) reduces to:

$$\left(\frac{n_m}{n_0}\right)_{t\to\infty} = \frac{\left(f_{pm} + f_{pi}\, f_{im}\right)\lambda_{0i}}{\left(f_{p0} + f_{pi}\, f_{i0}\right)\lambda_{md}}, \tag{47}$$

which is just what would be expected for the destruction of states 0 and m compared to their internal equilibration via state i. Nevertheless, if the internal equilibration rates are sufficiently rapid, then the following is obtained:

$$\left(\frac{n_m}{n_0}\right) = \frac{f_{im}\,\lambda_{0i}}{f_{i0}\,\lambda_{mi}} = \left(\frac{\lambda_{im}}{\lambda_{mi}}\right)\left(\frac{\lambda_{0i}}{\lambda_{i0}}\right) = \frac{g_m}{g_0}\exp\left(-E_m / kT\right), \tag{48}$$

which is the usual thermal-equilibrium population ratio.

In order to examine how rapidly the internal transition rates operate to produce the thermal equilibrium ratio in equation (48), $\lambda_{0d} = \lambda_{md} = \lambda_{id} = 0$ can be set and a re-examination of the polynomial $C(s)$ can be made to find that $C(s) \to sR(s)$, where $R(s)$ is now the quadratic polynomial $R(s) \equiv s^2 + \Lambda s + \Omega$, with $\Lambda \equiv \lambda_{i0} + \lambda_{0i} + \lambda_{im} + \lambda_{mi}$, and $\Omega \equiv \lambda_{mi}\lambda_{i0} + \lambda_{i0}(\lambda_{im} + \lambda_{mi})$. The two roots s_1 and s_2 of $R(s)$ are: $s_1 = [-\Lambda - (\Lambda^2 - 4\Omega)^{1/2}]/2$ and $s_2 = [-\Lambda + (\Lambda^2 - 4\Omega)^{1/2}]/2$.

To obtain the solution when there are no nuclear reduction rates we can simply set $\lambda_{0d} = \lambda_{md} = \lambda_{id} = 0$ in equation (42), and the calculations are similar to those achieved earlier.

4.4.3. General calculation of the approach to equilibrium

In the generalisation of the equilibrium, there arises the question how to handle the problem transitions between an isomer of a nucleus and its ground state when there is an ensemble of intermediate states involved that can all potentially couple with each other as well as with the isomer and the ground state. For the general calculation of long-lived isomers in stable nuclei (see Section 6) the formalism of Ward and Fowler [80Wa] can be used. Their work can be used on the problem of thermalization of ^{26}Al which has a long life. The coupled set of first order differential equations for the rates of change of abundance (N) of the levels of ^{26}Al can be found in matrix form as:

$$\frac{d}{dt}N = TN + \gamma\,\lambda_{p\lambda}\left({}^{25}Mg\right) - DN, \tag{49}$$

where the initial populating downward gamma cascades have been taken to the ground and isomeric states in order to be essentially instantaneous with the initial conditions: $N(t = 0) = 0$.

$$\gamma = \begin{bmatrix} 0 \\ \vdots \\ 0 \\ 1 - f_0\left({}^{26}\text{Al}\right) \\ f_0\left({}^{26}\text{Al}\right) \end{bmatrix}. \tag{50}$$

Additionally, the diagonal destruction matrix D contains all of the positron-emission, electron-capture, and nuclear reaction rates for leakage out of each level of ^{26}Al , i. e.:

$$D_{ii} = \lambda_{\varepsilon}(i) + \lambda_{\beta^+}(i) + \lambda_{p\gamma}(i) + \lambda_{np}(i) + \ldots, \text{ etc.}, \tag{51}$$

for state i. For simplicity the production of ^{26}Al would be restricted to the reaction of ^{25}Mg$(p, \gamma)^{26}$Al . Since the elements of N are subscripted sequentially in line with the decreasing excitation energy ending with the ground state, the elements of T are simply:

$$T_{ji} = \lambda_{ji}^* \quad \text{for } i < j; \quad T_{ij} = \frac{g_j}{g_i} \exp\left[-\left(E_j - E_i\right)/kT\right] T_{ji},$$

$$\text{for } i < j; \quad T_{ii} = -\sum_{j \neq i} T_{ji} \quad \text{for all } j, \tag{52}$$

where λ_{ji}^* is the stellar decay rate of state j down to state I, and the subscripts on the excitation energies E and on the statistical spin factors $g = 2J + 1$ indicate those for each appropriate state.

The total enhancement of a given internal spontaneous radiative decay rate, λ_γ, from a higher to a lower state is now given by:

$$\lambda^* = \lambda_\gamma \left(\left[1 - \exp\left(-E/kT\right)\right]^{-1} + \sum_l \alpha_l P_l + \gamma_{\text{elec}} + \Gamma_{\text{Coul}} + \Gamma_{\text{nucl}} \right), \tag{53}$$

where the connecting subscripts i, j, have been dropped for brevity. The first term of equation (53) incorporates the stimulated enhancement due to the surrounding stellar photon bath. The next term gives the usual enhancement observed in the laboratory due to bound-free internal conversion, where α_l is the internal conversion coefficient for the lth shell of bound atomic electrons; P_l is the fraction of the normal quota of l-shell electrons that are still bound in the stellar interior. The remaining enhancement factors – Γ – take into account the possible de-excitation caused by inelastic collision with the hot stellar plasma.

5. Status of isomer-excitations in the realisation of the γ-ray laser

The development of lasers that would generate radiation at still shorter wavelengths, comparable with atomic dimensions, can likewise be expected to lead to a vast array of new scientific and technological developments. The difficulty of graser developments, together with the perceived rewards of success, make it a major scientific and technological challenge. The problems of the graser are interdisciplinary: they involve nuclear and optical spectroscopy, chemistry, solid state physics and metallurgy. The subjects blend both basic science and technology. Three features common to all lasers are (1) an active medium (2) a host, and (3) a pump. The active medium is an ensemble of identical quantum systems (atoms, ions, or molecules) that undergo the lasing transition. The host serves to confine the active medium to a definite volume and geometry (CO gas in a glass tube, Cr^{+++} in a sapphire crystal, and Nd^{+++} in a glass rod). The pump supplies energy to excite the active medium to an inverted condition (in a similar way to the flash lamp for an Nd-glass and ruby laser, or an electron gun for CO_2 laser). In connection with the photoexcitation of the isomers of stable nuclei, only problems with the pumping of the graser are dealt with here.

5.1 Earlier concepts of the graser

The original proposal for a gamma-ray laser was made by Lev Rivlin [61Ri]. Two years later a strong momentum developed in relation to this topic as a result of earlier independent publications from the USA [63Ba], [63Va] and Russia [62Ka], [63Ch], [63Ri]. These first open publications illustrated the concerns that the applications might include military uses. Therefore, between 1965 and 1972, no reports of a new attack on the graser problem appeared in publicly available literature. All the early studies envisaged a chain reaction of induced transitions in radiochemically isolated isomeric nuclei having long lives, for which the pumping power is reduced to practically attainable levels. In principle, any radiation that can interact with a parent nuclide to form an isomeric state might serve as a pump. The following pumping radiation has been suggested: slow neutrons [63Ri], fast neutrons (I. Preiss, private communication from [81Ba]),

Fig. 23. The photograph was taken on the occasion of the Conference banquet (Proceedings of the ICPRA) on 29 March 1973. On the left is Edward Teller, in the middle is the author and on the right-hand side is Maurice Goldhaber

bremsstrahlung [63Ma], [73Bo], characteristic X-radiation [73Le], [79Vy], charged particle [74By], resonant (Mössbauer) radiation [64Ka], optical laser radiation [77Ri], [79Ar] and synchrotron radiation [74Di]. In the first cycle (which lasted 1963–1980) considerable attention was given to the problem of the sudden assembling of a critical density of pre-pumped nuclei to reach the threshold for stimulated emission.

After a number of investigations of the available possibilities there arose the dilemma of the graser. In other words: unfortunately, properties that should enable amplification, and those that would permit obtaining nuclear inversion densities sufficient for lasing are not compatible. The other problem involves the time required to activate, separate, concentrate and crystallise an appreciable number of isomeric nuclei by radiochemistry. It can require a few seconds, but more commonly it requires several hours. If inversion is to persist for a such long time, the transition lifetime must be appreciably longer. The Mössbauer effect, on the other hand, is presently observable only in the transition of a lifetime shorter than 10 μs. The intense pumping needed to invert isomers with lifetimes in the Mössbauer range may overheat, damage, or even destroy the solid that must support recoilless emission.

At the "International Conference on Photonuclear Reactions and Applications" (26–30 March 1973, Asilomar, USA) a "Panel Discussion of the Photonuclear Future" was organised (Moderator: E. Hayward; Panellists: P. Axel, W. Bertozzi, F. W. K. Firk, M. Goldhaber, W. Greiner, and E. Teller). Teller [73Te] spoke about the gamma laser and he said: "One topic about which there has not been a word said here is the gamma-ray laser." ... "A somewhat modest step in that direction would be to excite a nuclear state of ^{235}U (one of my favourite substances) which has an excited state of 60 volts, and a nuclear emission is a gamma emission even though it is 60 volts." After a short speech about the UV laser he said: "...I should not to be surprised that we might have – even ten years from now – a gamma-ray laser. This is a short-term prediction and therefore dangerous, because I might be still alive to see it happen or not see it happen." Despite this prediction and numerous proposals, attempts which have extended over nearly four decades have still not yet succeeded in developing such a graser. The conversations which covered this topic of photoexcitation are remembered in a photograph presented here *(Fig. 23)*.

At the end of this period it was generally accepted that such "brute force" approaches, usually requiring *in situ* pumping and involving only a single photon, were essentially hopeless. Baldwin, Solem and Gol'danskii [81Ba] published in the *Review of Modern Physics* an excellent summary paper titled "Approaches to the development of gamma-ray lasers". The scope of the authors made efforts to develop stimulated emission devices, using the Mössbauer effect in nuclear transitions, in order to generate coherent electromagnetic (EM) radiation in the 6 to 120 keV photon energy range. It was their hope that, by documenting what has been proposed and explaining why each proposal had fallen short of success, they would encourage others to contribute new ideas, to seek essential information now lacking, and to avoid the duplication of the past efforts; in this way the elusive goal might be reached. This review effectively documented the failure of traditional approaches to a graser – namely, those relying on the use of intense particle fluxes for input energy. Despite all this, they investigated many proposals and determined the minimal requirements of the applications by direct neutron and two-stage neutron pumping, and they also described numerous experiments of the separation of isomers.

Baldwin et al. [81Ba] investigated also the probable features of grasers. Geometrically any graser would necessarily require a crystal host, cooled to the lowest practicable temperature, since recoilless emission seems to be an essential feature. The majority of proposals have assumed the acicular form typical of a single-pass amplified-spontaneous-emission (ASE) device, since this ensures the access of pump radiation, the rapid removal of heat, and the escape of spontaneous radiation and the by-products of pumping, as well as a well-defined output beam. The main alternatives were: (a) closing the photon path by total external reflection, (b) achieving a closed photon path by a sequence of Bragg reflections, (c) reducing the non-resonant absorption by photon channelling in crystals, (d) suppressing all but a small number of Bragg modes in a perfect crystal, (e) bending a whisker crystal into a circle to achieve a closed photon path.

In the following part of this section some calculations will be presented and also some approaches – particularly those utilising techniques which can be realised in the pumping developments.

First, it is necessary to estimate the minimum number of excited nuclei that, in elongated filament geometry, would produce an appreciable effect of stimulated emission. Allen and Peters [71Al] used a condition that one net photon, in excess of those lost by non-resonant absorption and diffraction, could be stimulated by a photon as it passes from one end of the laser body to the other. This condition established a minimum inversion density N^* for a given length (L):

$$N^* > (N_0\sigma_a + L)/\sigma_s \tag{54}$$

the total concentration N_0 being for an undiluted graser isotope (the coefficient of L^{-1} should be ln 2 = 0.693). For a cylindrical body, the radius-to-length ratio is governed by the requirement of negligible diffraction loss:

$$F \equiv a^2 / L\Lambda >> 1. \tag{55}$$

The Fresnel number F should be at least three. In practice, the radius of the filament will be its limiting dimension. Thus a minimum volume of the active material should be at least:

$$V_{\min} = F\pi L^2\Lambda = \pi / a^4 / F\Lambda \tag{56}$$

and it will contain $N_2V_{\min}$ excited nuclei. If the degree of inversion is written as $N^* / N_0 = (N_2 - wN_1) / N_0 = p$, then the excited-state concentration is $N_2 = N_0 (w - p) / (w + 1)$, where w is the statistical weight ratio of the two states.

Combining these with the σ_s-stimulating cross-section and the ratio of Γ_r radiative width and total width Γ, the minimum number of excited nuclei is:

$$N_2V_{\min} = \left(\frac{2\pi a^2}{\Lambda^2}\right)\left(\frac{w+p}{pw+p}\right)\left(\frac{(1+\alpha)\Gamma T}{\beta f}\right)\left(1+\frac{N_0\sigma_a}{F\Lambda}\right). \tag{57}$$

The second factor accounts for the less-than-complete inversion, and the third one for both homogeneous and inhomogeneous line-broadening. The final factor gives the required excess of one photon over the number lost by non-resonant absorption. If the transition is between two excited states, then the effective lifetime is $T = T_1T_2 / (T_1 + T_2)$, and the branching ratio β may be less than unity.

The task of the pumping process does not simply involve the production of an adequate number of excited states but their production in high enough concentration in a solid line; this then allows a recoilless line. The total energy represented by even so small a total number of excited states was impressive.

5.2. Present status of graser research

Since 1980, the research of the feasibility of a graser has taken a completely new character. Numerous experiments ([81Ol], [84Co], [84De], [85Co], [88Ho], [89An]) have verified that the concept of quantum electronics can be applied at a nuclear level. In the "Proceedings of the 1st International Gamma-Ray Laser Workshop" (GARALAS'95) Collins and Carroll [97Co1] present an excellent lecture titled "Progress in the pumping of a gamma-ray laser". There are four unique advantages of the graser that accrue from its operation upon EM transitions of nuclei:

1. The constant linking of the frequency with the lifetime is more favourable according to equ. (1) by three orders of magnitude because of the accessibility of a variety of transition moments;
2. The isomer states can store energy in keV and even MeV for years;
3. Nuclear transitions can be routinely obtained. Without broadening, electromagnetic cross-sections are large and values for 1 nm transitions typically exceed the cross-sections for the stimulation of Nd in YAG; and
4. Working isomeric states can be concentrated in solid densities.

Those participating in the workshop dealt with many technical and theoretical problems involved in the development of devices analogous to the laser. For instance, strong interest in different kinds of studies were presented, such as: progress in the pumping of the graser [97Co1]; the prospect of lasing without inversion [97Ko]; the problem of the lasing without inversion due to a cooling subsystem [97Sh]; innovative approaches about the possibilities for γ-ray-stimulated emission experiments [97Ka]; the heat problem and the realisation of the pumping of gamma-ray solid lasers [97Ka]; and the production, separation, and preparation of isomers [97Ar], [97Be], [97Ca], [97Co2], [97Ho] [97Og].

Solem, in a paper [98So] dedicated to E. Teller on the occasion of his 90th birthday, also outlined the recent status of research on grasers. He made a short evaluation of some problems in the field of γ-ray emission in solids and estimated the performance of a graser system and the pumping (the latter is a central problem of γ-ray laser design).

According to Solem, the application possibility of photons for *directly-pumped* grasers, owing to the relatively low cross-section, is intrinsically inefficient (the number of isomers produced by photoexcitation and neutron-capture, and the stored-up energy density data are shown in *Table 9*). Except for bound-bound X-radiation, all readily available sources of photons and the energy needed lie within a rather broad band, calling for much higher power than estimated above. The Mössbauer effect greatly enhances the interaction of radiation with resonant nuclei, and the Bormann effect reduces the parasitic non-resonant absorption and scattering losses. The intense heating by nuclear recoil and the ionisation that would be created within the graser solid by photoelectric and Compton processes is devastating for M&B effects.

The *indirectly pumped* grasers offer the best hope for obtaining an inversion without destroying the conditions for M&B effects. Schemes for indirect pumping used to be divided into two-step, and two-stage categories.

In two-step pumping, the first step is to *create* a long-lived isomer by neutron reactor or, using a γ-irradiation facility, to *separate and concentrate* the storage isomer by radiochemical methods and thus *incorporate* the storage isomer in the host material of the graser body. The second step is to *quickly convert* the storage isomer into a short-lived upper lasing state, with corresponding homogeneous broadening to defeat inhomogeneous effects. The investigated methods were as follows:

a) Pumping with coherent radiation [86Ba], [86So];
b) Inter-level transfer by coherent radiation [73Mo], [79Iz], [81Co];
c) Inter-level transfer and pumping with incoherent radiation [73Le], [82Ol], Collins and Carroll [95Co] describe a hypothetical scenario in which a diamond host – 0.67 μm thick doped to 10% with an isomeric parent for 100 keV lasing transition – is pumped to the K-mixing level by absorption of 30 keV *bremsstrahlung*. With extremely optimistic parameters, they estimate that an energy fluency of 0.1777J cm^{-2} eV^{-1} within the resonant bandwidth should suffice for gain, and they conclude that the resultant temperature rise would not exceed the 2250 K Debye temperature of the host. According to the opinion of Solem [98So] these conditions seem unrealistic because: "…(1) at such a high degree of doping, the Bormann effect could not occur; (2) the effective Debye temperature would be only 590 K; and (3) the mechanical shock from delivery of multi-joule transfer pulse cannot be neglected; nor can the recoils from the 2 MeV γ-ray cascade to the lasing level".

In two-stage pumping, thermal neutron capture takes place in a blanket surrounding the graser body [73Go].

The next method is γ-ray lasing without inversion. This method has been demonstrated in optical laser research [84Sc], [92Sc], [95Ko] by techniques that used destructive interference. These established opposite phases for two substates of one of the lasing levels by means of a four-level scheme. If a workable scheme could be found for eliminating resonant absorption, the threshold excitation level for gain would be determined by the non-resonant scattering and absorption. Excitation requirements are still non-trivial, but might be reduced by as much as three orders of magnitude. If the anomalous transparency of the Bormann modes could be utilised, the excitation requirements could be reduced by about five degrees of magnitude. The applicability of the method was investigated using selection rules in a polarised population [79Vy], induced hyperfine transitions [79Ba1], and a level-crossing prospect [95Cou].

5.3. Evaluation of the stored energy-density data

In this section an attempt will be made to accumulate those data of photoexcitation of isomers which may be important in the developments of a graser. Thus, the determination of the maximum number of isomer-produced cases of photoexcitation by the irradiation of monoenergetic (Compton-scattered) radiation sources from the experimental cross-section data, and the calculation of the effective energy density of the graser host represent very important information. If it is supposed that the most favourable conceivable cases are available with regard to the saturation factor, the branching ratios, and the attainable enrichment of the abundance of the nucleus, then the number of isomers at the end of irradiation is:

$$N_f = \sigma_i N_i \phi_0 n' a, \tag{58}$$

where $N_f = I_k \times t_{1/2} / \ln 2$ and I_{exp} is the isomer activity (disintegration per sec), N_i is the initial number of nuclei in the target, σ_i (cm^2 eV) is the integral cross-section, Φ_o (= 1×10^{14} photons cm^{-2}) of the shot shown for the ^{60}Co radiation

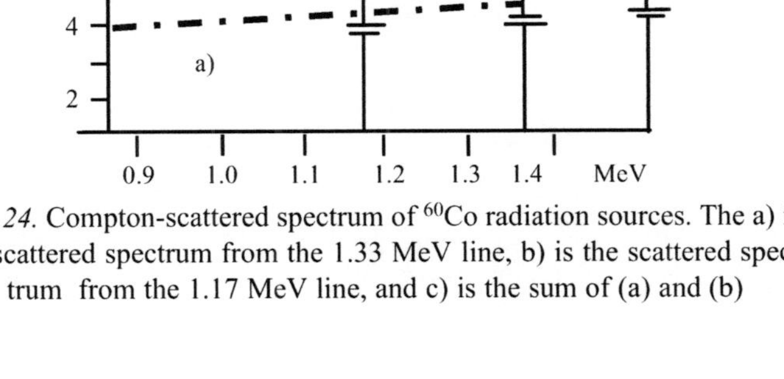

Fig. 24. Compton-scattered spectrum of ^{60}Co radiation sources. The a) is the scattered spectrum from the 1.33 MeV line, b) is the scattered spectrum from the 1.17 MeV line, and c) is the sum of (a) and (b)

sources, n' (eV^{-1}) is the number of scattered photons, and a is the natural abundance of the nucleus in the target.

Scattering from the energy hν to the $h\nu' + \Delta\nu'$ resonance range is calculated by the Klein–Nishina formula, valid for Compton-scattering by atomic electrons which are free in this energy range. This n' value is shown in *Fig. 24.* The experimental conditions are similar to those applied by Veres [63Ve1,2]. The number of n' scattered into an 1 eV energy interval per γ-quantum from 0.25 MeV to 1.3 MeV energy range is calculated at $\phi_0 = 1\times10^{14}$ γ-quanta cm^{-2} s^{-1} flux of the source. In some cases when there is a low number of activation levels in the decay scheme (for example, in the case of ^{77}Se the partial decay scheme contains 10 activation levels), the calculation becomes more complicated.

Carroll et al. [91Ca1,2] published the photoexcitation of isomers using *bremsstrahlung* sources having 4 and 6 MeV endpoint energies (E_0). The time-integrated yield of the number of isomeric states, and N_f obtained by irradiating a number of nuclei N_i in ground state with photon flux Φ_0 in photons/cm^2, delivered a *bremsstrahlung* continuum of intensities up to an end point energy is:

$$N_f = N_i \Phi_0 \int_0^{E_0} \sigma(E) F(E,E_0)\, dE, \tag{59}$$

where $F(E, E_0)$ is the intensity distribution within the *bremsstrahlung* spectrum, normalised such that:

$$\int_0^{E_0} F(E,E_0)\, dE = 1 \tag{60}$$

and $\sigma(E)$ is the cross-section for the (γ, γ') reaction. N_i and N_f are the populations of initial and final nuclear states, respectively. Then the integrated cross-section for $E_a = 4$ MeV is:

$$(\sigma\Gamma) = \frac{N_f}{N_i \Phi_0 F(E,E_0)}, \tag{61}$$

where $F(E, E_0)$ are shown in *Fig. 25* at $E_0 = 4$ MeV and $E_0 = 6$ MeV end energies. For the DNA/PHITON linac shot shown, $\Phi_0 = 4\times10^{14}$ photon cm^{-2} and for the DNA/AURORA the linac shot $\Phi_0 = 4\times10^{13}$ photon cm^{-2}.

Based on the experimental cross-section data of equations (58)–(59) an estimation is shown here of the number of N_f (see column 8 of *Table 9*). Supposing optimal conditions for triggering the nucleus from the isomeric state to an excited level with a very short lifetime (τ^* is μsec or shorter), which lies above the isomeric state, the storage energy densities of the graser host are tabulated in column 9 of *Table 9.*

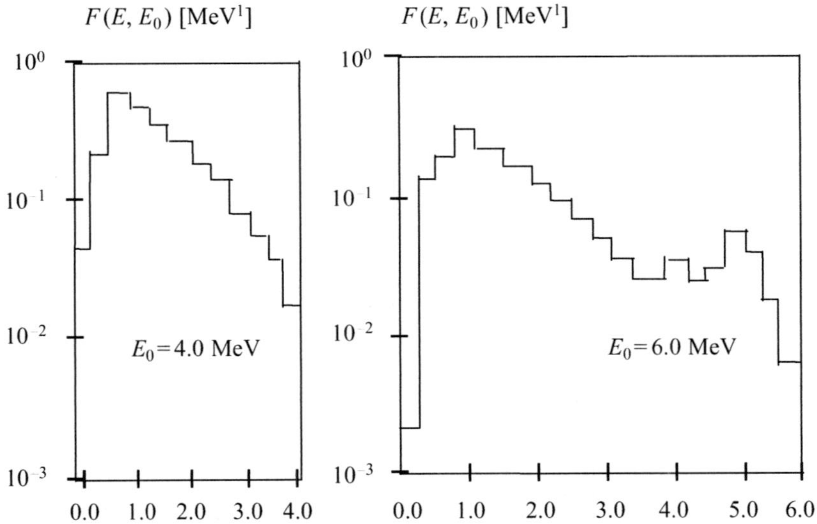

Fig. 25. Relative spectral intensities $F(E, E_0)$ of *bremsstrahlung* produced by Linacs used to irradiate target materials [91Ca2]

From the data in *Table 9* it can be seen that the isomers may be natural sources of high energy densities. (The energy densities displayed in column 8 refer to the related energies during the lasing transition. However, the energy needs for pumping, should also be considered.)

The quantum mechanics teaches that the motions of the charged particle found within the nucleus will represent the highest velocities of circulation possible in any material. The high-energy densities, great velocities and the charges usually reradiate such energies in a time which is too short to be measured ($<10^{-18}$ s). In rare cases, however, selection rules sufficiently inhibit the coupling of the particle motion to the electromagnetic fields for the energies to be stored for a long time (even for thousand years with respect to certain special nuclei).

In *Fig. 26* transition probabilities are shown as a function of energy for all multipolarities of interest for the special conditions: $A = 100$; $S = 1$ (it has been assumed that $a = 1.2 \times 10^{-13} A^{1/3}$ cm and $\mu_P = 2.79$).

One idea is to stimulate the isomeric state with ΔE trigger energy to an upper state which has a very short lifetime, then a considerable part of the storage energy of the isomeric state is transferred to the upper excited level, from which the storage energy is suddenly released *(Fig. 27)*.

Recently the amplification of lasing without inversion attracted much attention. The main idea is to suppress the resonant absorption via atomic interference in three-level atoms. Hence this opportunity to release a graser

Table 9. The available energy densities in a graser host, calculated for the maximum available intensity

Isomer	E_m a)	E_{st} a)	ΔE a)	$t_{m,\text{‰}}$ b)	$\tau(ps)$ c)	$N_f(g^{-1})$	E_d(J/g) d)	E_{dh}(W/g) e)	$J^{\pi}_{\Delta E}$
1	2	3	4	5	6	7	8	9	10
$^{77}Se^m$	162	239	77	17.4 s	25	1.4×10^{12}	1.8×10^{9}	255	M2 (E3)
$^{79}Br^m$	207.1	217.1	10	4.9 s	34	7×10^{10}	6.9×10^{7}	8	M2 (E3)
$^{87}Sr^m$	388.5	1253.9	865.4	2.8 h	2.8	2.2×10^{11}	4.9×10^{8}	455	M2 (E3)
$^{89}Y^m$	909	1744.6	835.6	16.1 s	0.5	5×10^{9}	1.4×10^{9}	28	M2 (E3)
$^{99}Tc^m$ f)	142.7	181.1	38.4	6 h	3610	2.2×10^{15}	1.4×10^{10}	5247	M2 (E3)
$^{103}Rh^m$	39.8	357.4	301.4	56 m	74	3.6×10^{11}	3.4×10^{8}	155	E1 (M2)
$^{107}Ag^m$	93.1	324.8	231.7	44.3 s	5	1.9×10^{11}	5.5×10^{8}	210	M2 (E3)
$^{109}Ag^m$	88.0	311.4	223.4	39.6 s	5.9	2.9×10^{11}	6.9×10^{8}	260	M2 (E3)
$^{111}Cd^m$	396.2	416.7	20.5	49 m	120	9×10^{11}	4.8×10^{6}	1.5	M2 (E3)
$^{113}Cd^m$ f)	263.6	316.2	52.6	14.1 y	1.1×10^{4}	3.1×10^{19} g)	1.2×10^{14}	3.8×10^{7}	E3 (M4)
$^{113}In^m$	391.7	1024.3	632.6	1.66 h	3.6	1.4×10^{12}	2.5×10^{10}	6576	M2 (E3)
$^{115}In^m$	336.2	941.4	605.2	4.5 h	15.1	1.3×10^{12}	4.7×10^{9}	92	M2 (E3)
$^{117}Sn^m$	314.6	711.5	396.9	13.5 d	980	5.5×10^{13}	2.8×10^{9}	745	M2 (E3)
$^{119}Sn^m$ f)	89.5	787.0	697.5	293.1 d	190	1.3×10^{13}	9.8×10^{9}	257	M2 (E3)
$^{123}Te^m$	247.6	489.8	242.2	119.7 d	3.1×10^{4}	4.6×10^{12}	5.9×10^{6}	1.3	M2 (E3)
$^{125}Te^m$	144.8	636.1	491.3	57.4 d	40	1.1×10^{12}	5.1×10^{8}	114	E2
$^{129}Xe^m$	236.1	321.7	85.6	8.9 d	44	2.5×10^{11}	2.1×10^{8}	3.6	E3 (M4)
$^{131}Xe^m$	163.9	637	473.1	11.8 d	5.6	1.2×10^{11}	5.6×10^{8}	9.2 h)	M2 (E3)
$^{135}Ba^m$	268.2	874.5	606.3	28.7 d	1.3	1.1×10^{12}	3.5×10^{10}	4425	M2 (E3)
$^{137}Ba^m$	661.7	1252.5	590.8	2.6 m	0.4	1.2×10^{11}	3.2×10^{10}	6416	M2 (E3)
$^{167}Er^m$	207.8	346.6	138.8	2.27 s	1000	1.4×10^{12}	4.7×10^{7}	15	E2
$^{176}Lu^m$	123.0			3.6 h		3.4×10^{12}	3.4×10^{12}		
$^{178}Hf^{m1}$ f)	1147.4			4 s	7×10^{4}	1.6×10^{12}	4.2×10^{6}		
$^{178}Hf^{m2}$ f)	2446	~3000	~0.5	31 y	7×10^{4}	1.8×10^{12} i)	4.0×10^{14}	4.8	
$^{179}Hf^{m1}$ f)	375	518.3	123.3	18.7 s	200	4.8×10^{14}	1.5×10^{11}	7×10^{4}	E2 (M3)
$^{179}Hf^{m2}$	1105.8			25.05 d	600 j)	1.2×10^{12}	3.7×10^{8}	176	
$^{180}Hf^m$ f)	1145.5	1472.3	326.8	5.5 h	140 j)	3.9×10^{15}	5.1×10^{12}	2.4×10^{6}	
$^{180}Ta^g$ x)	75.3		2800	8.1 h	9.4	3.3×10^{12}	4.2×10^{9}	2525	
$^{183}W^m$	309.5	453	143.5	5.15 s	8500	2.3×10^{15}	6.1×10^{9}	4225	M2 (E3)
$^{189}Os^m$	30.8	233.5	202.7	5.8 h	280	2.0×10^{13}	5.7×10^{7}	43	E2 (M1)
$^{191}Ir^m$	171.3	343.2	171.9	4.94 s	20	7.3×10^{11}	1.0×10^{9}	808	M2 (E3)
$^{193}Ir^m$	80.2	357.7	277.5	10.5 d	19.8	2.9×10^{11}	2.0×10^{8}	158	M2 (E3)
$^{195}Pt^m$	259.3	562.8	397.9	4.02 d	14	2.9×10^{11}	8.6×10^{8}	665	M2 (E3)
$^{197}Au^m$	409.2	547.5	138.3	7.73 s	4.6	1.1×10^{11}	1.6×10^{9}	1100	M2 (E3)
$^{199}Hg^m$	532.5			42.6 m		9.9×10^{10}			

The letters shown below can be found in the table and explanations for their respective meanings are given here:
a) E_m-isomeric state, E_{st} – the presumed stimulation level, and $^*E$ – the estimated triggering photon energy in keV; b) half-life of isomeric state; c) the lifetime of the presumed stimulation level; d) E_d – the energy density; e) E_{dh} – energy density of the graser host; f) produced by $(n, ^*)$ reaction; g) irradiation: 1 year; h) specific weight is 0.46 g cm^{-3} at pressure of 40 atm; i) the maximum number of nuclei $^{178}Hf^{m2}$ is 4.8×10^{15}, produced in a $^{176}Yb(\alpha, 2n)$ reaction, irradiated by U-200 cyclotron (Dubna); j) Weisskopf estimate; x) the ground state has a short half-life of 8.1 hours, and the half-life of the 75.3 keV isomeric state is $>1.2\times10^{15}$ years

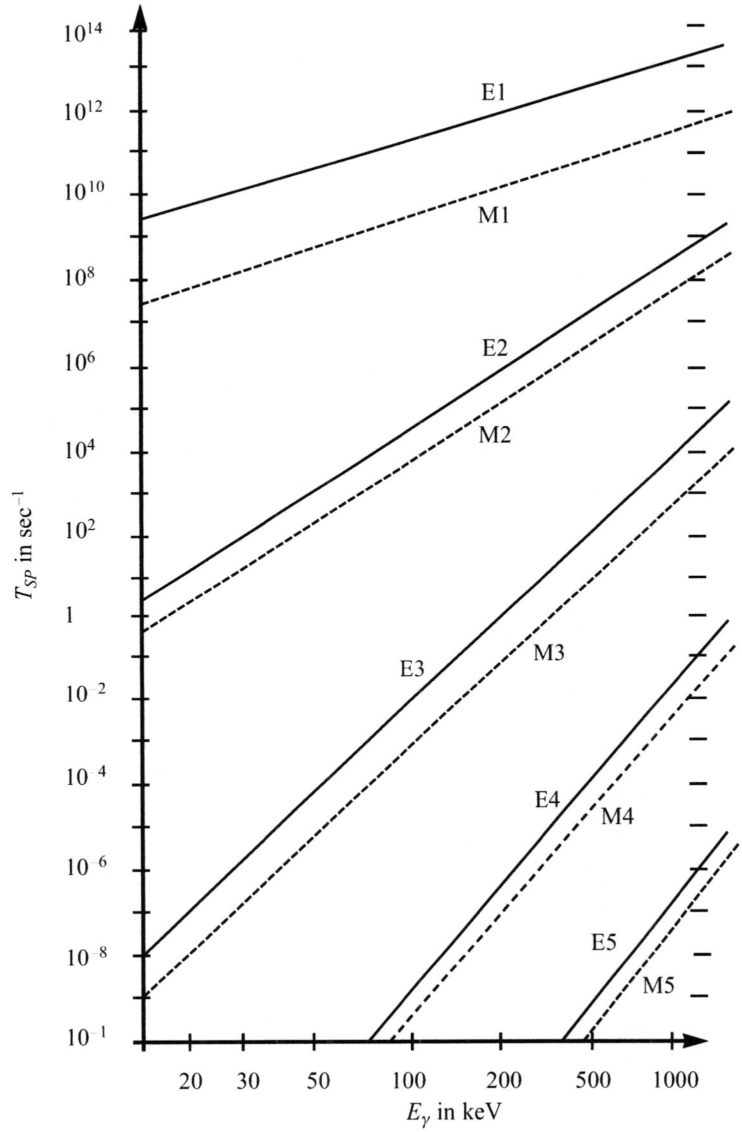

Fig. 26. Gamma-decay transition probability for a single photon [eqs (62) and (63)] as a function of γ-ray energy for various multipolarities

without the need for population inversion on the basis of the coherent and interference effects deserves the most careful study. The process can be described as $\sigma_{mi} = f\,\sigma_{exci}$, where $f = (E_m \to E_{exc}) / (E_{exc} \to E_g)$. The f depends on the quantum numbers (spin/parity), the transition energy, and the multipolarity – i.e., $f = f_s f_E f_L$. In the calculation the formalism of Moszkovszki is used from his work on the theory of multipole radiation [79Mo]. It is convenient to classify electromagnetic radiation by multipole orders L, according to the angular momentum carried off by each quantum. For each multipole order there are two possible classes of radiation: electric 2^L pole (EL), and magnetic 2^L pole (ML); these differ with respect to parity.

The conversion of angular momentum and parity of the system of nucleus and γ-rays imposes selection rules on the possible multipolarities of a γ-transition: between two states of specified angular momenta (J_i , J_f) and parities (π_I, π_f). $|J_i - J_f| \le L \le |J_i + J_f$ and $\Delta\pi \equiv \pi_I / \pi_f = (-1)^L$ for EL radiation, and $\Delta\pi \equiv \pi_I / \pi_f = (-1)^{L-1}$ for ML radiation. Here $\Delta\pi = +1$ denotes no parity change and $\Delta\pi = -1$ denotes parity change.

The gamma-decay transition for a single proton is:

$$T^{(\mathrm{EL})} = \frac{4.4(L+1)}{L[(2L+1)!!]^2}\left(\frac{3}{L+3}\right)^2\left(\frac{\eta\omega}{197\ \mathrm{MeV}}\right)^{2L+1}\times$$
$$\times\left(a, \text{in } 10^{-13}\ \mathrm{cm}\right)^{2L} S\left(j_i, L, j_f\right)\times 10^{21}\ \mathrm{sec}^{-1}, \tag{62}$$

$$T^{(\mathrm{ML})} = \frac{0.19(L+1)}{L[(2L+1)!!]^2}\left(\frac{3}{L+2}\right)^2\left(\mu_P L - \frac{L}{L+1}\right)^2\left(\frac{\eta\omega}{197\ \mathrm{MeV}}\right)^{2L+1}\times$$
$$\times\left(a,\ \text{in } 10^{-13}\ \mathrm{cm}\right)^{2L-2} S\left(j_i,\ L,\ j_f\right)\times 10^{21}\ \mathrm{sec}^{-1}. \tag{63}$$

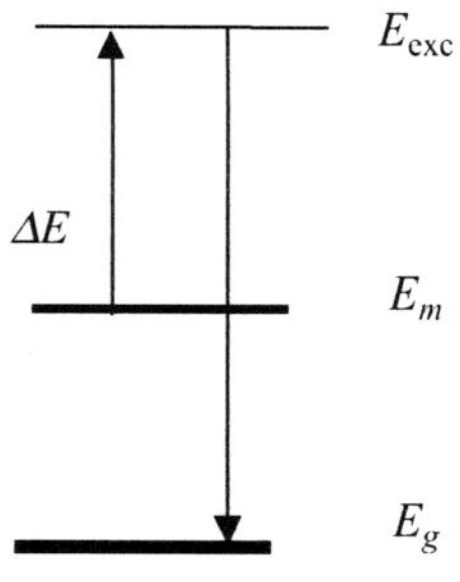

Fig. 27. A simple scheme of inversionless lasing from the isomeric state

It is also of interest to consider the electromagnetic transitions of a single neutron. Due to the absence of electric charge, one would expect no interaction between orbital motions of neutrons and the electromagnetic field. However, the orbital motion of the rest of the nucleus around the centre of mass of the whole nucleus will give a matrix element for electric transitions, which is smaller than the matrix element for a single proton in the corresponding transitions by a factor Z/A^L. This is correct for E1 transitions, and the transition rate for E1 radiation. However, for higher multipoles, there are additional corrections associated with the centre of mass. It has been shown [55El] that for harmonic oscillator wave-functions, the reduction factor is not $(Z/A^L)^2$ but 0 for neutrons, and 1 for protons. An additional but also small contribution to the electric multipole matrix element comes from the transition of neutron spin with the electromagnetic field. According to a strict single-particle picture, magnetic transition probabilities in odd-neutron nuclei are expected to be slightly smaller than the corresponding magnetic transition probabilities in odd-proton nuclei $[\mu_N / (\mu_P - - (L+1)^{-1})]^2$.

In those configurations with many particles the calculations for gamma-ray transition probabilities can be expected to be more realistic than those with several particles. It is often convenient to express the results in units of single-particle transition probabilities and to define a reduced transition probability as follows:

$$B(\sigma L, J_i \to J_f) = \frac{1}{2J_i + 1} \sum_{m_i} \sum_{m_f} \sum_{M} |< f | \sum M_{LM}^{\sigma}(p) | i >|^2. \tag{64}$$

The transition probability, which depends on the detailed nuclear structure only through B, is then given by:

$$T_{i \to f}^{(\sigma L)} = \frac{8\pi(L+1)}{L[(2L+1)!!]^2} \frac{1}{\eta} \left(\frac{\omega}{c}\right)^{2L+1} B(\sigma L, J_i \to J_f). \tag{65}$$

To evaluate this it is convenient to define a statistical factor S, as follows:

$$S(j_i, L, j_f) = 4\pi (2j_i + 1)^{-1} \sum_{m_i} \sum_{m_f} \sum_{M} \left| \int_{4\pi} \Theta_{l_f, j_f}^{mt*} Y_{LM}^{*} \Theta_{l_i, j_i}^{m_i} \, d\Omega \right|^2. \tag{66}$$

A simple expression for S can be derived:

$$S(j_i, L, j_f) = (2j_f + 1)\left[C(j_i \, j_f \, L, \, ½ - ½ 0)\right]^2, \tag{67}$$

where C is a Clebsch–Gordan coefficient.

For a given j_i, L, j_f, of the four possible kinds of transition, ($l_i = j_i \pm 1/2$, $l_f = j_f \pm 1/2$), two have the parity change $(-1)^L$ required for an electromagnetic, EL radiation – that is to say, for those transitions for which $l_i - l_f - L$ is an even integer. The expression for S takes a comparatively simple form for the special case that $| j_i - j_f | = L$. For $j_f = j_i - L$, $l_f = l_i - L$, we have:

$$S(j_i,\, L,\, j_f) = \frac{(j_i - ½)!}{(2j_i)!!} \times \frac{(2L+1)!!}{L!} \times \frac{(2j_f)}{(j_f - ½)!}. \tag{68}$$

In particular, $S(L+1/2,\, L,\, 1/2) = 1$. Values of $S(j_i,\, L,\, j_f)$ are listed in *Table 10.*

Table 10. Coefficients $S(j_i, L, j_f)$ for $L = | j_i - j_f$

$j_f \downarrow$; $j_i \rightarrow$	1/2	3/2	5/2	7/2	9/2	11/2	13/2
1/2		1	1	1	1	1	1
3/2	2		6/5	9/7	4/3	15/11	18/13
5/2	3	9/5		9/7	10/7	50/33	225/143
7/2	4	18/7	12/7		4/3	50/33	700/429
9/2	5	10/3	50/21	5/3		15/11	225/143
11/2	6	45/11	100/33	25/11	18/11		18/13
13/2	7	63/13	525/143	1225/429	315/143	21/13	

Table 11. Estimates [equations (62) and (63)] for the transition probability of a single proton

Multipolarity	T_{SP} in sec^{-1}			Comment	
E1	1.0×10^{14}	$A^{2/3}$	E_γ^3	S	($a = 1.2\times A^{1/3}$
M1	2.9×10^{13}		E_γ^3	S	10^{-13} cm,
E2	7.4×10^{7}	$A^{4/3}$	E_γ^5	S	$\mu_P = 2.79$,
M2	8.4×10^{7}	$A^{2/3}$	E_γ^5	S	E_γ in MeV)
E3	3.4×10^{1}	A^{2}	E_γ^7	S	
M3	8.7×10^{1}	$A^{4/3}$	E_γ^7	S	
E4	1.1×10^{-5}	$A^{8/3}$	E_γ^9	S	
M4	4.8×10^{-5}	A^{2}	E_γ^9	S	
E5	2.5×10^{-12}	$A^{10/3}$	E_γ^{11}	S	
M5	1.7×10^{-11}	$A^{8/3}$	E_γ^{11}	S	

The estimates [equations (62)–(63)] for various multipolarities are listed in *Table 11*. Although the status of graser research is unresolved, there are a few areas where there is definite promise. The first one is in the suppression of resonant absorption and the possibility of γ-ray lasing. A possible development of science and technology would be in the following methods: (i) reduction of resonant absorption, (ii) elimination of parasitic scattering and non-resonant photon loss, (iii) limiting of inhomogeneous broadening, (iv) control of power dissipation and retention of the benefits that accrue to the solid state.

The next section deals with the following: a) the accumulation of the main results of photoexcitation published by different laboratories, b) the illustration of some special cases with regard to the utility of results in the nucleosynthesis and, c) calculation of the maximum energy densities accessible in a graser host.

6. Experimental and theoretical results of isomers of individual stable nuclei

In this section a compilation of published, measured, and some calculated data is given. A great deal of useful information has been acquired with respect to the electromagnetic interaction of nuclear charge and current. All stable nuclei having an isomeric state (and contained in *Table 4*) have been dealt with separately. Recently, many level-scheme derivations have been reported using this (γ, γ') method, including the determination of level parameters. The accumulated cross-section, quantum number and activation level energy, as well as numerous other data, also permit statistical studies of the level widths and fluctuation in the partial level widths of transitions to low-lying levels; furthermore, comparison can be made with the level properties in the well-known neutron region. While the basic method described in this book relies on random overlap between the resonance level and a discrete energy neutron line, several techniques exist for varying the energy of the γ line, thus increasing the range of levels accessible for study.

Several hundred resonance levels excited by the (γ, γ') reaction in a wide variety of stable nuclei have been discussed in the literature. The level schemes of the various target nuclei have been determined by analysis of the energy spectrum of the measured isomeric activity and the quantum numbers of the activation levels, as well as the isomeric and ground states. This analysis relies on certain assumptions regarding the nature of the γ transitions. Where the activation levels de-excite through two or more cascade transitions, it is assumed that the higher energy transition is the following one. It is assumed that the excited levels are usually the same in the analysis of decay schemes as in the (n, γ) and (n, n'γ) reactions, Coulomb excitations, and in other spectroscopy studies.

In many experiments involving photoactivation, the authors have usually determined the effective cross-sections from the isomeric activity measurements in those cases where the energy of the activation level and quantum numbers of the target nucleus were not known. In other cases those data were given by the integrated cross-sections, which contain more information for calculating the lifetime of the activation level and for the application of the result of (γ, γ') excitations in other studies. However, in the case of higher energy levels, secondary, third and even higher activation transition levels cannot be completely ruled out, especially where the two transition energies are close. Where more than one activation level is excited using a given (γ, γ') source, it is necessary to identify the inelastic transitions from each level. This can occasionally be done with a good choice of the energy of the radiation sources. Very often it is difficult to determine the energy of the first activation level, when the γ-line energy of the radiation sources are very different from each other. An individual example of a typical case will be discussed in the determination of the first activation level of ^{195}Pt, where two isotopic radiation sources (^{60}Co and ^{182}Ta) are used. The measured values of the cross-sections made it possible to identify the activation level by comparing them with previously measured results (for details see Section 6.25). In some cases (^{107}Ag and ^{109}Ag), when both the half-life and the energy values of the isomeric states are very close to each other, the real value can be identified using enriched isotopes.

In the tables of this section the most characteristic data of the isomer activation experiments have been compiled. First, there are the effective and integral cross-section data, derived from irradiation experiments by *bremsstrahlung*, monoenergetic radiation sources, positron sources and other irradiation facilities. In order to be able to distinguish between the integrated cross-section values obtained from irradiation by *bremsstrahlung* or by monoenergetic (Compton-scattered) γ radiation sources, different marks were used. Values marked by Σ_{int} in mb keV or μb MeV units are related to a group of activation states excited by *bremsstrahlung*, while σ_{int} values are evaluated in eV cm^2 for single activation levels from experiments involving monoenergetic sources.

Before turning to the study and presenting the experimental and calculated cross-section and other data of the isomers indicated in *Table 4,* it is worth mentioning the ^{26}Al nucleus which has a half-life of 7.2×10^5 years and is therefore absent from *Table 4.* However, there are observations in certain meteorites of ^{26}Mg isotopic abundance excess that are correlated with the Al/Mg elemental ratios. This helps to establish that ^{26}Al was present in the early solar system. The astrophysically short half-life implies that the nucleosynthetic events must have occurred no more than a few million years prior to the formation of these meteorites. Norman and Kellogg [85No2] investigated the processes for the production of ^{26}Al. At a sufficiently high temperature the ^{26}Al^g(γ, γ')^{26}Al^m reaction may occur and the isomer transmutes by ^{26}Al^m(n, p)^{26}Mg, or $^{26}\text{Al}^m \rightarrow {}^{26}\text{Mg} + \beta^+ + \nu_e$. The level structure of the nucleus introduces some complications to the study of ^{26}Al. The ^{26}Al ground state has 5^+ spin/parity and a half-life of 7.2×10^5 years. The ^{26}Al^m at 228 keV has 0^+ and β^+ decays to ^{26}Mg, with a halflife of 6.3 seconds, but does not decay to ^{26}Al^g. Thus any ^{26}Al formed in this isomeric level will not survive long enough to become incorporated into a meteorite as ^{26}Al.

After some consideration, in a few cases nuclei having isomeric states above 0.5 seconds were dealt with (although they are not found in the nature), because they contain valuable information about the study of the nucleus or may be important in recent applications (nucleosynthesis or gamma laser). On the other hand, discussion is avoided in those cases in which no sufficient information is available to draw any conclusions.

In what follows, photoexcitation experiments are discussed one by one for each element having an isomeric state.

6.1. Experimental and calculated data of ^{77}Se and its recent applications

The element selenium has six stable isotopes; among them only the isotope ^{77}Se has an isomeric state (energy: E_{iso} = 161.9 keV; half-life: $t_{1/2}$ = 17.4 sec). This nucleus occurs in nature with abundance of 7.63%. The quantum numbers of the ground state and the isomeric state are $1/2^-$ and $7/2^+$, respectively.

6.1.1. Accumulation of photoexcitation data of ^{77}Se

The main experimental and theoretical data published in the literature about the photoexcitation of ^{77}Se have been gathered here. A summary of the experimental results based on measured isomeric activities, published by different laboratories, are listed in *Table 12* along with the relevant physical parameters and references. On the basis of the cross-section values the following general remarks can be made:

- The activation cross-section values associated with the sources emitting quanta of higher energies demonstrate the existence of some activation levels. The differing experimental cross-section values for a given isomer measured with different radioactive γ-sources of slightly different γ-energies are due to resonance flux variations produced by Compton-scattering from the source material.
- Some information can be acquired on the first and the subsequent activation level energies from the γ-energy of the radioactive sources and from *bremsstrahlung* with endpoint energies that have been altered gradually.
- In the higher energy range, a group of activation levels is excited frequently and the integrated cross-section grows up strongly.

Table 12. The excitation cross-sections and other data of the ^{77}Se nucleus

Irradiation facility	Excitation E_{exc} (MeV)	σ_{eff} (nb)	Σ_{int} (mb eV)	$g\,\Gamma_0\,\Gamma_m/\Gamma$ (meV)	Life-time τ (ns or ps)	Refs
2 MeV electron	0.25±0.01		0.8±0.7/0.25	1.3×10^{-5}	9.7 ns $^{d)}$	[69Bo]
accelerator of	0.44±0.01		1.8±1.2/0.4	9.1×10^{-5}	23 ps $^{d)}$	
bremsstrahlung,	0.52±0.01		5.7±5/1.5	4.1×10^{-5}	4 ps $^{d)}$	
BS	0.825±0.01		4.6±4/1	8.0×10^{-4}	0.52 ps $^{d)}$	
	0.925±0.01		35±3/1	7.8×10^{-3}		
	1.05		32±3/1	8.4×10^{-3}	3.2 ps $^{d)}$	
	1.19		180±120/40	6.7×10^{-2}		
	1.60		520±40/10	3.5×10^{-2}		
0.2–1 MeV X-ray;	0.25		2 $^{e)}$			[87An]
20 ns imp.	0.48		15 $^{e)}$			
Marx-generator	0.818		7 $^{e)}$			
	1.005		300 $^{e)}$		0.36 ps	[88An]
Four accelerators,	4		$(3.3\pm0.4)\times10^3$			[91Ca2]
0.5–11 MeV; BS	6		$(6.5\pm0.32)\times10^4$			[91Ca2]
2–7 MeV range of the	1.7±0.1		$(1.5\pm0.15)\times10^3$			[93Ca1]
LINAC, Darmstadt	2.6±0.1		$(1.2\pm0.12)\times10^4$			
	3.2±0.1		$(3.7\pm0.37)\times10^4$			
	4.0±0.1		$(5.5\pm1)\times10^4$			
4–15 MeV; M-30	8		$(2.4\pm0.1)\times10^6$			[93Ma1]
Microtron, BS	10		$(3.5\pm0.5)\times10^6$			
4 MeV, Imp. Ac. BS	E>1 $^{a)}$	5100	3.7×10^4 $^{c)}$			[95La2]
15 TBq; ^{24}Na	<2.75 $^{c)}$	<400				[69Ab]
100 TBq; ^{24}Na	<2.75 $^{c)}$	490	6500±700			[73Ve2]
^{116}In^{m}; reactor loop $^{b)}$	<2.112 $^{c)}$	330±80				[69Ab]
3 TBq ^{142}Pr	<1.576 $^{c)}$	285				[73Ve2]
100 TBq ^{60}Co	1.1 $^{a)}$	95	560	0.175		[63Ve2]
280 TBq ^{60}Co	1.19 $^{a)}$		250	0.093		[73Za]
10 TBq ^{46}Sc	<1.121 $^{c)}$	14				[73Ve2]
65 TBq ^{137}Cs	0.5206		13.6±7.2			[95La1]
65 TBq ^{137}Cs	0.521		11.6±5.6		4 ps	[95Kh]

a) Estimated values; b) reactor loop; c) E_{max} of the radiation sources; d) from other experiments, [96Fi] $^{d)}$; e) $\Sigma_{int} = \pi b_a b_0 \Gamma \sigma_0/2$

The partial level diagram of ^{77}Se up to 1230 keV is given in *Fig. 28.* Transitions and levels relevant to the photoactivation are only indicated. The known major activation levels E_{Ai} can be excited by resonance energy gammas from continuous spectra of intense *bremsstrahlung* or isotopic γ-sources; the line spectra of the latter undergo large degradation by Compton-scattering in the real geometries employed. Relevant branching ratios for transitions leading to the isomeric state (either directly or through a cascade, and for the direct transi-

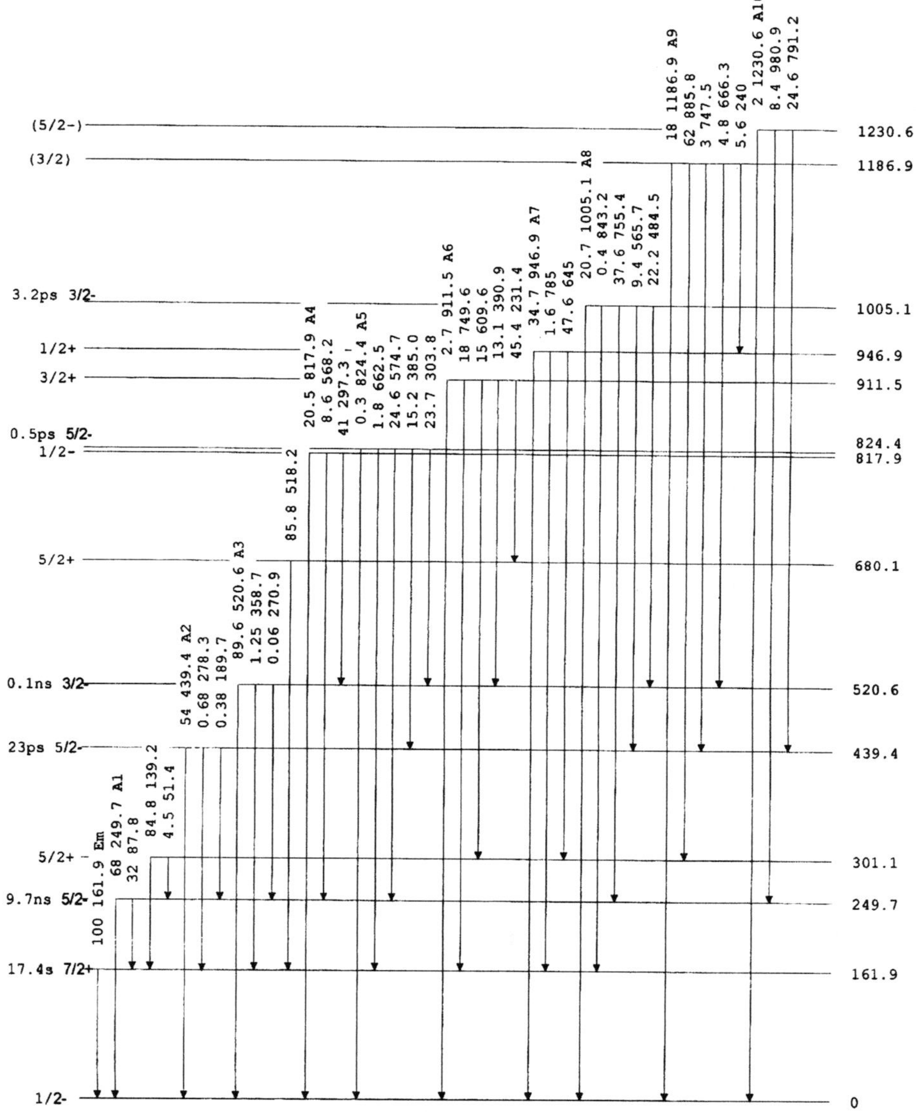

Fig. 28. The γ-excitation and γ-decay partial scheme of ^{77}Se. On the left side of the figure are the quantum numbers and the respective lifetimes of the levels, and the on right side are placed the energies of the levels in keV. On top of the levels are the branching ratios and transition energies; Em marks the isomeric (metastable) level and *A*1, *A*2, ..., *A*10 represent the activation levels. Some data are taken from [96Fi]

tion to the ground state) are indicated as percentages. As can be seen from *Fig. 28,* in this low (up to 1230 keV) energy region the ^{77}Se the nucleus has about ten excited levels, which were identified as activation levels.

Integrated isomer cross-sections were determined either from adopted nuclear data or from measured isomer yields and calculated resonance flux densities. Level lifetimes (taken from the literature) were also determined by other nuclear reactions, radioactive decays and Coulomb excitations, or from (γ, γ') reaction results. In the latter case, when an exceedingly large number of activation levels is present it is difficult to calculate the lifetime of these levels because the relevant data are mostly insufficient. If two activation levels are populated, the σ_{eff} and resonance cross-sections σ_{r1}, σ_{r2} are connected with each other as follows:

$$\sigma_{\text{eff}} = \frac{\sigma_{r1}}{1+a} + \frac{\sigma_{r2}}{1+a^{-1}} \quad \text{where} \quad a = \frac{N(E_2)\int \sigma_{r2}\, dE}{N(E_1)\int \sigma_{r1}\, dE} \qquad (68)$$

and $N(E_1)$ is the number of isomer of the first activation level E_1 and $N(E_2)$ is the number of the isomer of the second activation level E_2.

In the energy range of 2–50 MeV the model of the photoexcitation of isomeric nuclei (discussed in recent work) is shown in *Fig. 29.* Only two intermediate states are drawn as examples and the number of those actually participating depends upon the sequence of energy levels and transition probabilities for the particular nucleus being considered. With these excitation mechanisms Carroll et al. [91Ca2], and Sáfár et al. [91Sá] studied the systematics of giant resonances for the photoexcitation of isomeric states; this was in order to deter-

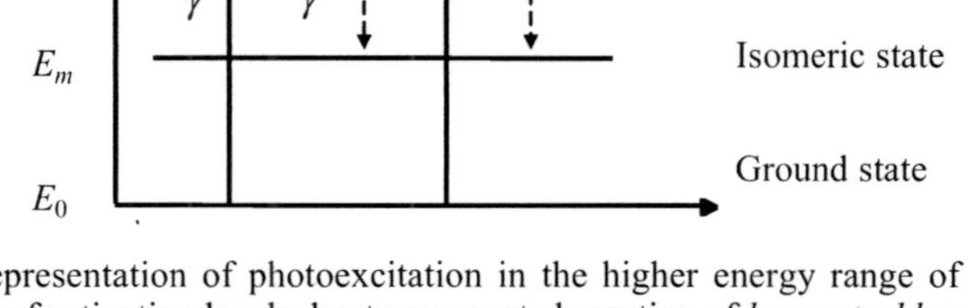

Fig. 29. Schematic representation of photoexcitation in the higher energy range of activation levels. The population of activation levels due to resonant absorption of *bremsstrahlung* X-rays is denoted as γ. Two clusters of activation levels are shown at excitation energies E_1 and E_2, but the actual number will vary for each nuclide. Populations partly decay to the final state either directly or as a part of a cascade of radiative transitions, as indicated by the dashed arrows

mine whether such results were curiosities associated with only a few isotopes or more generally prevalent phenomena. They [91Ca2] claimed that their results were a few orders of magnitude larger than those usually characterising $(\gamma, \gamma')^{m}$ reactions; even so, in view of the results for other nuclides this statement cannot be confirmed. A secondary objective was to learn whether the large integrated cross-sections were the results of a large number of reaction channels of more conventional size or rather, a few of unprecedented magnitude.

In the case of numerous activation levels it is also a possibility to study the intermediate structures of the nucleus [67Bo], [90Du], [90Po]. Among practical applications it is worth mentioning the importance of selenium as a trace element in animal fodder [77Ve]. The selenium deficiency leads to a disease characterised by symptoms such as "muscle white colour" and "exudative meat" (facts which have been proved experimentally, too). The minimum quantity which it is possible to determine is between 1 μg/g–15 μg/g [63Ka], [72Ka], [66Ve2], [69Ve], [78Ve2], [80Ve2].

6.1.2. Impact of the *(γ, γ)* reaction on nucleosynthesis in the Se Br Kr region

The first basic assumption of the *s*-process is that neutron-capture rates are slow compared to β-decay rates. The element krypton has two *s*-only isotopes, and it is apparent that only a fraction of the *s*-process current passes through ^{80}Kr, whereas all of it passes through ^{82}Kr. If the *p* abundance is ignored – which seems justifiable in light of the small abundance of ^{78}Kr and other *p* nuclei not pictured – the result is approximately $\sigma(^{80}\mathrm{Kr}) / \sigma(^{82}\mathrm{Kr}) \approx \lambda_{\beta}(^{79}\mathrm{Se}) / [\lambda_{\beta}(^{79}\mathrm{Se}) +$ $+ \lambda_{n}(^{79}\mathrm{Se})]$. On the basis of systematics it can be expected that $\sigma(^{80}\mathrm{Kr})$ will be about twice as great as $\sigma(^{82}\mathrm{Kr})$. This expectation can be coupled with the known abundance in order to estimate the branching ratio at ^{79}Se. An essential complication is that the β-decay lifetime of ^{79}Se is not a known constant; instead, it depends upon the temperature. The ground state of ^{79}Se is found in the laboratory to have a half-life equal to 6.5×10^{4} years, but it has an excited state of only 96 keV, and when the decay rate of that thermally populated excited state is considerable in the total rate, one finds that the half-life becomes $\tau_{1/2}(^{79}\mathrm{Se}) =$ $= 6.5 \times 10^{4}$ years $/ [1 + 6.47 \times 10^{5} \times \exp(-11.1/T_{8})]$; this has only 34 years at $T_{8} = 2 \times 10^{8}$ K°.

The ^{85}Kr plays an important role in the usual branching calculations [76Wa] because the usual competition between neutron-capture and β-decay can occur. From the discussion of Ward [77Wa], it can be seen that a 3-times larger neutron flux can now be allowed and this still gives the same total flow through

$^{86,87}Sr$ if the isomer branching ratio is sustained. Thus any *s*-process production of ^{86}Kr and ^{87}Rb would also be lowered significantly by isomer branching at ^{85}Kr. Since the cosmo-radiogenic $^{87}Rb \rightarrow {}^{87}Sr$ decay is an *s*-process chronometer, these considerations show that the use of this chronometer is much more complicated than indicated by Clayton [64Cl].

A more important requirement is the concordance of the *s*-process environments for simultaneously producing the correct branching ratios at ^{79}Se and ^{85}Kr. This is indicative of the mean thermodynamic conditions prevailing during the *s*-process system event.

In this mass region there are three isomers of stable nuclides – i.e., $^{77}Se^{m}$ (161.92 keV, with a half-life of 17.36 seconds), $^{79}Br^{m}$ (207.1 keV with a half-life of 4.86 seconds), and $^{83}Kr^{m}$ (41.543 keV with a half-life of 1.83 hours) as indicated in *Fig. 30* by an * mark. Each of these isotopes directly de-excites from the isomeric state to the ground state by short-half-lives. Thus there is no experiment suitable for studying similar problems, as in the case of ^{85}Kr in astrophysical circumstances.

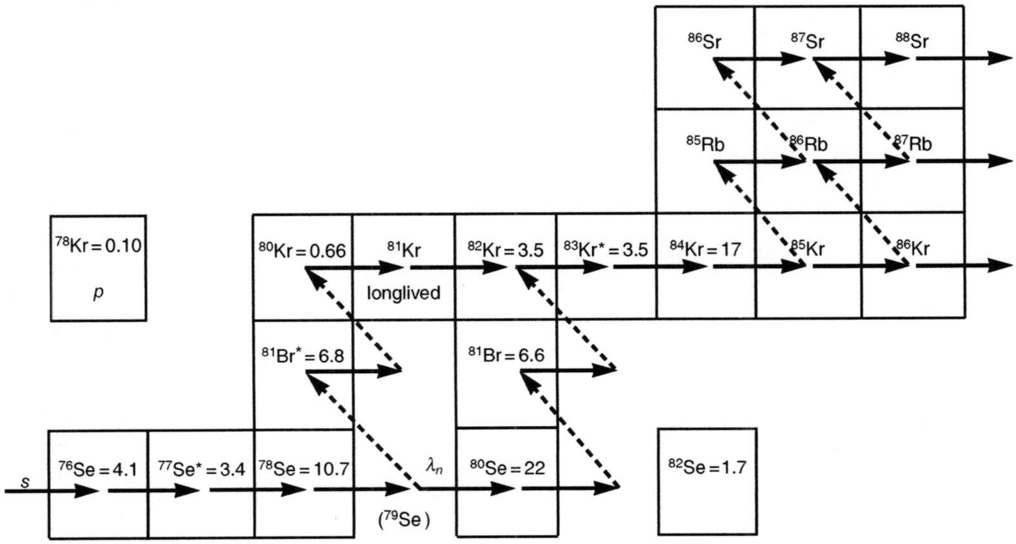

Fig. 30. The *s*-process path through selenium, bromine, krypton, rubidium and strontium. A branch between neutron capture and β-decay occurs at ^{79}Se ($T_{1/2,\text{lab}} = 6.5\times10^4$ years). The ^{80}Kr and ^{82}Kr are shielded from *r*-process production by ^{80}Se and ^{82}Se, respectively. The abundance of each nucleus per 10^6 silicon atoms in the solar system is indicated. The * marked isotopes have an isomeric state

6.1.3. Applicability of ^{77}Se for a graser pump

The integral cross-sections evaluated from experiments by Compton-scattered monoenergetic radiation source (^{60}Co), and by LINAC *bremsstrahlung* are shown in *Table 12.* From these the average values are:

a) $\sigma_{i,av}$ = 210 mb eV at 1.2 MeV endpoint energy, measured by Boivin [69Bo], Veres [63Ve2] and Zaparov [73Za] as 180, 200, and 250 mb eV, respectively.
b) From 1.57 MeV to 10.0 MeV endpoint energy, the integral cross-section values ([μb MeV], [mb MeV], [μb keV], [μb MeV] or [10^{-29} cm^2 keV]) are indicated in the equivalent [mb eV] unit.

These results are in agreement with each other within error bars, except for the works of Carroll et al. [91Ca1,2], [93Ca1], where the cross-sections are determined as $(330\pm4)\times10^{-29}$ cm^2 keV and $(5500\pm1000)\times10^{-29}$ cm^2 keV, respectively, at 4 MeV endpoint energy. The next very high energy density in a graser host is the consequence of the high integral cross-section (2.4±0.1 mb MeV at 8 MeV and 3.5 ± 0.5 mb MeV at 10 MeV endpoint energy) of Mazur et al. [93Ma1].

Numerical calculations were performed with the integral cross-section values (mb keV) given in the second column of *Table 13.* It was supposed that the

Table 13. The energy density

E_{ep} (MeV)	Σ_i(mb eV)	N_f (10^{12})	E_d (GJ/g)	E_1 (W/g)	[Ref.]
1.2	180	4	4.1	706	[69Bo]
1.2	200	4.3	4.5	784	[63Ve2]
1.2	250	5.5	5.7	980	[73Za]
1.57	850	19	19	3334	[73Ve2]
1.6	520	11	12	2040	[69Bo]
1.7	1500	33	34	5883	[93Ca]
2.6	12000	260	270	4.7×10^4	[93Ca]
2.75	4900	110	110	1.9×10^4	[73Ve2]
3.2	37000	810	840	1.5×10^5	[93Ca]
4.0	3300	72	75	1.3×10^4	[91Ca2]
4.0	55000	1200	1250	2.2×10^5	[93Ca1]
4.0	37000	810	840	1.5×10^3	[95La2]
6.0	65000	1400	1500	2.5×10^5	[91Ca2]
8.0	2400000	53000	54000	9.4×10^6	[93Ma1]
10.0	3500000	77000	79000	1.4×10^7	[93Ma1]

nucleus from the isomeric state (E_m = 162 keV energy) could be stimulated to move up to an excited level of 250 keV and a lifetime of 25 ps, with 77 keV trigger-energy; this excited level decays very quickly (25×10^{-12} sec) to the ground state. The material of a graser host should contain enriched ^{77}Se of 90%; the number of isomers N_f (g^{-1}), the energy densities (Jg^{-1}), and the power densities (wg^{-1}) of the graser host (~0.18 μg) are tabulated in the third, fourth, and fifth columns of *Table 13*. The multipolarity of triggered photons in the low energy range may be $M2$ ($E3$), as can be seen from *Table 9*. Nevertheless, it is questionable whether, in the high energy region, the conditions of the stimulated transitions will be improved or not. (The increasing energy needed for triggering to higher energy levels are not considered here.)

6.2. ^{79}Br

The element bromine (Z = 35) has two stable isotopes. One is ^{79}Br with 50.69% natural abundance and the other one is ^{81}Br (49.31%). The isomeric state of ^{79}Br has a 4.9 second half-life and the energy, spin and parity of this state are 217.1 keV and $9/2^+$, respectively. Since 1969 a number of papers have dealt with its production by *bremsstrahlung* of electron accelerators, and γ-rays of radioactive sources [69Ve], [73Za], [73Ve2], [78Ve1], [87An], [88An], [88Ga2], [91Ca2], [93Ca1], [95Kh], [95La1], [95La2].

6.2.1. Photoexcitation data

The experimental activation level energies, cross-sections, level-widths and lifetime values of ^{79}Br, measured by different laboratories, are summarised in *Table 14*.

From the data of *Table 14* it can be seen that before 1987 only rough estimates were available for the value of the activation level energy.

Anderson and Collins [87An], [88An] described a new technique for the absolute calibration of intense sources of pulsed radiation in the 0.2–1 MeV range of photon energies. The scheme of the model used to calibrate the *bremsstrahlung* spectrum is shown in *Fig. 4*. Most of the selenium contribution occurs at the 480 keV line (actually two lines at 420 keV and 521 keV), and all the contributions from ^{79}Br occur at 761 keV. The γ-ray activation technique is based upon the exposure of samples of the irradiating spectrum in the form of populations of nuclei excited to isomeric states with lifetimes ranging from seconds to hours.

Table 14. Isomer excitation cross-sections and other data of the ^{79}Br nucleus

Irradiation facility	Excitation E_{exc} (MeV)	σ_{eff} (nb)	σ_{int} (mb eV)	Σ_{int} (μb keV)	$g\Gamma_0\,\Gamma_m/\Gamma$ (meV)	Lifetime τ (ns)	Refs
3 TBq ^{46}Sc	<1.12	6			4.3×10^{-3}		[73Ve2]
6 PBq ^{60}Co	1.12–1.33	10.7			8.6×10^{-3}		[73Ve2]
30 TBq ^{142}Pr	<1.57	57					[73Ve2]
98 TBq ^{24}Na	<2.75	65					[73Ve2]
280 TBq ^{60}Co			40				[73Za]
1 MeV X-ray	0.761			62			[88An]
Four ACC.	4			6600±170			[91Ca2]
0.5–11 MeV, BS	6			19000±230			[91Ca2]
1.5–7 MeV	1.8±0.1			650±60			[93Ca1]
ACC, BS	3.2±0.1			8500±900			
	4.4±0.2			70000±15000			
4–15 MeV, BS	8			0.84±0.12			[93Ma1]
	10			3.1±05 (mb MeV)			
4 MeV LINAC	4	660		5.9 (μb MeV)			[95La2]
65 TBq ^{137}Cs	0.384		3.0±1.0			10.8	[95Kh]

Calibration of the *bremsstrahlung* spectrum required some initial assumptions about its form. The high-energy portion of this curve is the linear Kulenkampff approximation [67Ku] with the endpoint energy. At a certain energy it joins a region of constant intensity that extends to a cut-off level. The cut-off level was chosen to be 200 keV since all of the pump lines are considered to lay above this value.

The knee in the spectrum was placed at 350 keV. Inserting the experimentally obtained isomer populations and the cross-sections for Se and Br from *Tables 12* and *14,* the solution for the endpoint energy is 920 keV. Thus, the solution given above most realistically describes the intensity in the interval between these lines (^{77}Se "line" at 480 keV and ^{79}Br line at 761 keV).

Of primary importance is the fact that substantial populations of isomeric states can be produced by pulsed *bremsstrahlung* irradiation with relatively modest intensity sources. This technique has been exploited for the development of an important new method for characterising pulsed, high-energy X-ray devices.

From these results – and using the data of [96Fi] and from the measurement of Lakosi et al. [95Kh] – it is possible to construct the most relevant partial level scheme up to 761 keV *(Fig. 31).*

In the higher energy range (2–15 MeV) Carroll et al. [91Ca2], [93Ca1], Lakosi et al. [95La2], Mazur et al. [93Ma1,2], and Veres et al. [73Ve2] have

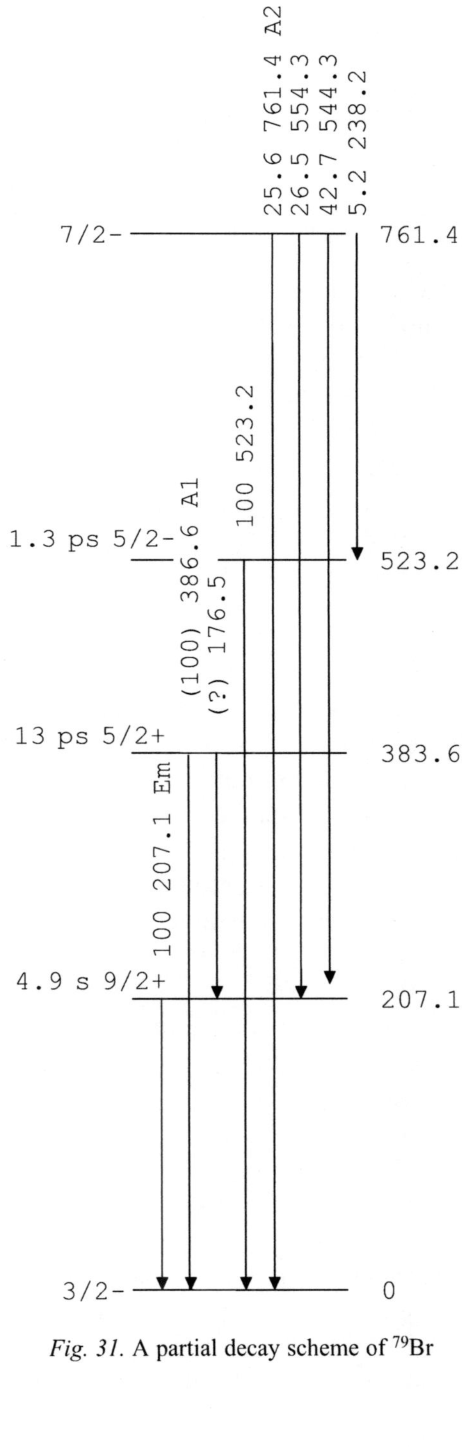

Fig. 31. A partial decay scheme of ^{79}Br

published cross-sections and other values produced by *bremsstrahlung* and radioactive sources.

Carroll et al. [91Ca2] examined the photoexcitation of 19 nuclides – among them ^{77}Se and ^{79}Br. Furthermore, they draw another conclusion that, with a few obvious exceptions, the general trend supports the hypothesis that the principal nuclear structure responsible for the order-of-magnitude enhancement of (γ, γ') reactions lies below 4 MeV. Much smaller increases occur between 4 MeV and 6 MeV and perhaps none between 6 MeV and 10 MeV. There is no persuasive evidence that additional channels open between 6 and 10 MeV. In turn this seems to imply that the giant dipole resonance has nothing to do with the enhancement of these (γ, γ') reactions.

Ponomarev et al. [90Po] studied the anomaly in the energy dependence of the yield of the isomer in ^{81}Br by photoexcitation, as a manifestation of an intermediate nuclear structure. The isomeric yield was normalised by the fixed number of accelerated electrons stopped in the betatron target by a tantalum sheet 1.5 mm thick – i.e. slightly thicker than the range of 4 MeV electrons. A considerable increase in the *bremsstrahlung* produced a yield of the isomer ^{81}Br with $J_m^\pi = 9/2^+$, E_m =536 keV and $t_{m,1/2}$ = 36 μs was observed as a function of the endpoint energy in the range 3–4 MeV. The activation level with E_{ex} = 3.18 MeV(5/2$^+$) was obtained in the framework of the quasi-particle phonon model. This has a wave-function with a 13% contribution of the quasi-particle component $2d_{5/2}$, and an 83% contribution of the quasi-particle phonon component [$1g_{9/2}2_1^+$]; it provides the resonance in the total scattering cross-section and is well linked to the isomeric state. From the experiment it was concluded that the first component is the low-lying fragment of the quasi-particle state $2d_{5/2}$. The valence transition of the proton from the ground state is the reason for the strong radiative excitation of this level (with a value $B(E1\uparrow) = 0.03\ e^2 fm^2$). It would be interesting to obtain information from the direct excitation of the level with $J^\pi = 5/2^+$, E_{ex} = 3.45 ± 0.15 MeV and Γ_0 ~ 0.7 eV in the reactions of elastic photon scattering at a low momentum transferred.

No useful data exist on the implication of the isomer activation of ^{79}Br^m in graser technical developments.

6.3. ^{83}Kr

The element with the atomic number 36 is the gaseous state krypton, which has six stable isotopes, among them the ^{83}Kr (natural abundance: 11.5%); this has an isomeric level with a half-life of 1.83 hours. The quantum numbers of this nucleus for the ground and the isomeric states are $J^\pi = 9/2^+$, and $1/2^-$, respectively, and the energy of the isomeric state is 41.5 keV. The level structure of ^{83}Kr was studied in the radioactive decay of ^{83}Br and ^{83}Rb and in (^{3}He, α) reactions by Lederer and Shirley [78Le]. Using the $^{80}(\alpha, n\gamma)^{83}$Kr reaction Eskola et al. [82Es] inferred spin-parity assignments. In the following an overview is given on the production of ^{83}Kr^{m} by photoactivation, showing that the isomer production via (γ, γ') proceeds by populating the activation levels which feed the metastable level of ^{83}Kr.

6.3.1. Experiments using the (γ, γ) reaction

The ^{83}Kr^{m} was produced by direct X-ray bombardment of the element in 1945 by Wiedenbeck [45Wi]. The γ-rays emitted in the transition from the metastable level to the ground state have been found to be 29 and 46 keV, respectively. Counters with brass cathodes of 2.5 cm diameter were constructed and filled with a mixture of krypton gas to a pressure of 20 cm of Hg and ethyl ether to a pressure 2 cm of Hg. After bombarding this counter with intense X-rays of a maximum energy of 2.5 MeV, moderate activity could be observed; this then decayed, having a half-life of about 115 minutes.

For the production of ^{83}Kr^{m} via (γ, γ') reaction by irradiation of γ-rays of ^{60}Co source the only experiment was carried out by Tam et al. [91Ta]. They have shown that the isomer production proceeds by populating the activation levels lying up to the 1170 keV level. An indirect method was used for determining the integral cross-section of ^{83}Kr^{m} by making comparison with the ^{115}In^{m} excitation implemented under similar conditions. Using this value, the integral cross-section for the ^{83}Kr(γ, γ')^{83}Kr^{m} reaction at the 1170 keV activation level was 32±7 mb eV. On calculating the ratio $\sigma_{\text{int}}(Kr)/\sigma_{\text{int}}(In)$ an assessment was made for the ratio of the resonance flux densities at the activation energies of ^{83}Kr and ^{115}In. The latter is known to be 1.078 MeV. By looking at the partial level scheme of ^{83}Kr *(Fig. 32)* it was possible to calculate the integral cross-section for the individual levels using the Breit–Wigner formula. In this way an integral cross-section of 29.5 mb eV was obtained at the 1170.4 keV level. For the lower lying activation levels (562 and 798 keV) the sum of three cross-sections less than 0.5 mb eV (i.e., 1.5%) was calculated. This means that 1170.4 keV can be considered as the first major activation level of ^{83}Kr. Because the ^{60}Co radiation source has 1173.24 keV and 1332.52 keV monoenergetic γ-lines, the Compton-scattered γ'-flux produced by both gammas of the irradiation source can populate the isomeric state of ^{83}Kr. In this case the experimental cross-section happens to be 32±7 mb eV, which is in good agreement with the value calculated from the level scheme.

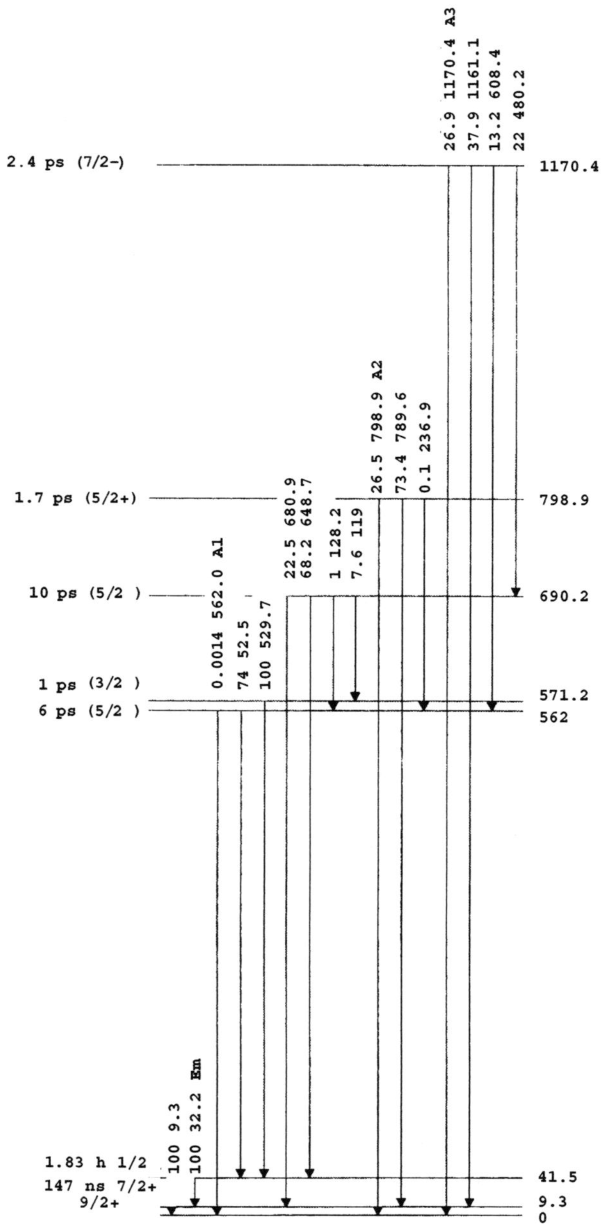

Fig. 32. A partial level scheme of ^{83}Kr

From level data – i.e., transition γ-line energies, branching ratios, and quantum numbers of the levels taken from the Nuclear Data Sheets [86Mü] and from the measurement [91Ta] – it was possible to construct the most probable partial excitation and de-excitation scheme, as shown in *Fig. 32.* In the figure the level and transition energies are in keV, and the branching ratios indicated after the transition energies are given as percentages.

Four years later another experiment was also carried out using a 4 MeV linear accelerator and *bremsstrahlung* from a 0.9 mm Pt converter placed against the electron beam of 4 MeV, with a 25 μA beam-current [95La2]. It was found that the effective cross-section for a flux >1.3 MeV is 1.3 μb. Furthermore, the authors excited a series of isomers and the isomer activation cross-sections integrated up to 4 MeV were also determined. This was done using the Brink–Axel hypothesis and the Lorentzian extrapolation of the photoabsorption cross-sections for a reference given for giant dipole resonance [88Di]. The results show a large variety in the respective magnitudes of the cross-sections. Large cross-sections have been obtained and in these the spin difference (ΔI) between ground and isomeric states is relatively small, though a large strength is also displayed for ^{176}Lu, in spite of the fact that $\Delta I = 6$. The particularly strong excitations for deformed nuclei are striking. Extremely low cross-sections may be due to low level densities in half-magic nuclei and high-lying isomeric levels. Strong excitations are expected where the isomer is accessible from a level populated by $E1$ transition from the ground state. The results demonstrated that isomer activation is a multiparametrical process.

6.3.2. Implications of krypton isomers

For the implication of the isomer activation of ^{83}Kr^{m} in graser technical developments we have no useful data.

Numerous investigators have examined the various path and mechanism by which the isomeric states of nuclei that are formed during nucleosynthesis are brought into statistical thermal equilibrium with their ground states. For example, Ward [77Wa] made a re-examination of the effect of non-thermal isomeric branching for many key *s*-process isotopes. Attention is focused on the effects of isomeric states if the stellar *s*-process environment is incapable of quickly thermalizing their initial non-equilibrium populations. These isomeric states are significantly populated by either the γ cascades following neutron capture on the preceding isotope or directly by the β-decay of an unstable isobar. For the study of the importance of long-lived isomers in the *s*-process branching he investigated some isomer among them ^{83}Kr^{m}, too. To asses the effects on the

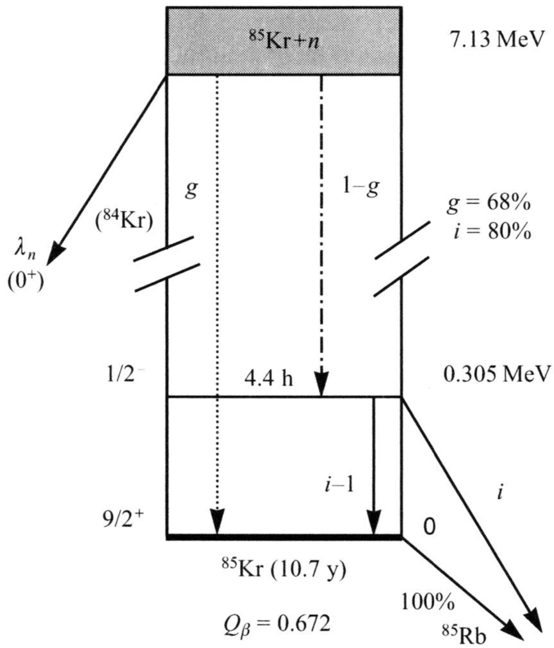

Fig. 33. A scheme of *n*-capture and isomer branching of ^{85}Kr

usual *s*-process he has illustrated the prototype situation by the branch-point example of ^{85}Kr for an excited state whose β-decay rate can successfully compete with its spontaneous radiative decay rate down to the ground state. *Fig. 33* schematically shows the detailed neutron capture and isomer-branching scheme for ^{85}Kr.

It has been assumed that a fraction, g, of the neutron-capture via ^{84}Kr leads to the 10.7 y ground state, while the remaining part, $1-g$, of the neutron-captures results in direct population of the 4.4 h isomeric state. A further assumption is that the characteristics of the isomeric state are such that a fraction, h, β-decays directly to ^{85}Rb , and, therefore, that $1-h$ goes into the ground state. We will retain g and h as variables fore the general case, although at thermal energies (0.0025 eV), neutron-capture by ^{84}Kr results in $g = 0.68$, and the isomeric state at 305 keV β-decays with $h = 0.80$. If these branching ratios survive the temperatures and densities of the stellar environment, then the *s*-process flow equations describing the abundances of the ground state of ^{85}Kr and its stable daughter ^{85}Rb should be modified to read:

$$\frac{1}{\sigma\left({}^{85}\mathrm{Kr}\right)}\frac{d\Psi\left({}^{85}\mathrm{Kr}\right)}{d\tau}=(1-gh)\,\Psi\left({}^{84}\mathrm{Kr}\right)-\left[1-f_{-}\left({}^{85}\mathrm{Kr}\right)\right]^{-1}\Psi\left({}^{85}\mathrm{Kr}\right), \tag{69}$$

and

$$\frac{1}{\sigma\left({}^{85}\mathrm{Rb}\right)}\frac{d\Psi\left({}^{85}\mathrm{Rb}\right)}{d\tau}=\left[\frac{1}{f_{-}\left({}^{85}\mathrm{Kr}\right)}\right]^{-1}\Psi\left({}^{85}\mathrm{Kr}\right)+gh\Psi\left({}^{84}\mathrm{Kr}\right)-\Psi\left({}^{85}\mathrm{Rb}\right), \tag{70}$$

where $f_{-}(^{85}\mathrm{Kr})$ is the branching ratio resulting from the competition between neutron and β^{-} emission at the ground state of ^{85}Kr [76Wa].

In the astrophysical setting for the *s*-process $kT \approx 30$ keV is given, thus the actual value of *g* under stellar conditions will depend on the level density characteristics within 30 keV of the $^{84}\mathrm{Kr} + n$ separation energy (7.3 MeV) in the compound ^{85}Kr nucleus. For a typical heavy nucleus with $A \approx 120$ level density is at least of the order of a hundred levels per keV, which increases with the increasing atomic weight. In such a case, if stellar de-excitation of the state is slow compared with the time required for the state to capture a neutron, then some of the *s*-process flow could proceed via the isomeric state to the next isotope; the flow would partially bypass the next isobar and the subsequent portions of the alternate β-decay path. Since the abundance of the longlived isomer would not be negligible, its decay after the *s*-process event would have to be included as a contributor to the abundance of its stable daughter.

For steady flow through the branch system – with $f_{-}(^{86}\mathrm{Rb}) = 1$ – the combination of isomer branching and the usual competitive branching at the ground state of ^{85}Kr would affect the remaining species as:

$$\Psi\left({}^{86}\mathrm{Sr}\right)=\Psi\left({}^{87}\mathrm{Sr}\right)=\left[gh+f_{-}\left({}^{85}\mathrm{Kr}\right)(1-gh)\right]\Psi\left({}^{84}\mathrm{Kr}\right), \tag{71}$$

and

$$\Psi\left({}^{86}\mathrm{Kr}\right)=\Psi\left({}^{87}\mathrm{Rb}\right)=\left[1-f_{-}\left({}^{85}\mathrm{Kr}\right)\right](1-gh)\,\Psi\left({}^{84}\mathrm{Kr}\right). \tag{72}$$

The general extension of these results to other branching networks is easily obtained by appropriate modifications to the above results.

Since the cosmo-radiogenic $^{87}\mathrm{Rb} \rightarrow {}^{87}\mathrm{Sr}$ decay is potentially an *s*-process chronometer, these considerations – coupled with the effects of an intermittent neutron flux and the small cross-sections in this region – show that the use of this chronometer is much more complicated than the discussion of Clayton [64Cl] indicates. (The abundance of ^{87}Rb is approximately equal to the amount of ^{87}Sr produced by nucleosynthesis.) Concordance of the *s*-process environments for simultaneously producing the correct branching ratios at ^{79}Se and ^{85}Kr, as indicative of the mean thermodynamic conditions prevailing during the solar system *s*-process event, requires to consider the unthermalized isomeric flows through ^{85}Kr.

6.4. ^{87}Sr

The element strontium has four (^{84}Sr, ^{86}Sr, ^{87}Sr and ^{88}Sr) stable nuclei and their natural abundance are 0.56%, 9.86%, 7%, and 82.58%, respectively. Among them, the ^{87}Sr nucleus has an isomeric state with a half-life of 2.8 hours and level energy of 385.5 keV. The quantum numbers of the isomeric and ground states are $J_m = 1/2^-$ and $J_g = 9/2^+$, respectively.

6.4.1. Experimental and theoretical results

Wiedenbeck [45Wi] excited first the metastable state of ^{87}Sr by the direct X-ray bombardment of strontium in a (γ, γ') reaction and by the electron by bombardment of strontium chloride in (*e*, *e*') reaction, using a Van de Graaff electrostatic generator. However, he gave no information about the cross-section of this reaction and on the lowest energy state of excitation levels investigated. Boot and Brownson [67Bo] excited ^{87}Sr by photons and electrons at energies below 3 MeV. The excitation function is shown in *Figure 34.* The most probable activation

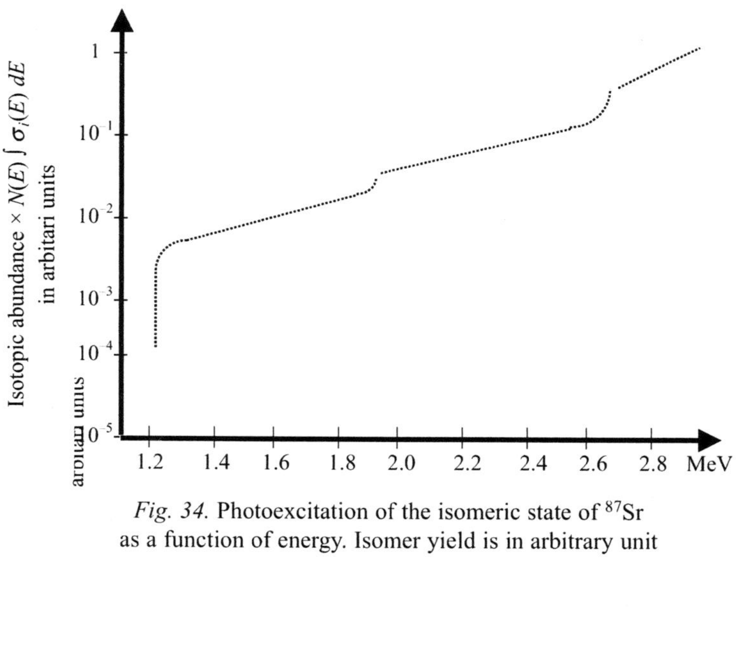

Fig. 34. Photoexcitation of the isomeric state of ^{87}Sr as a function of energy. Isomer yield is in arbitrary unit

levels are shown in the figure. Only three activation levels (1.22, 1.88 and 2.66 MeV) were determined from measurement data, while an activation level may be at $E_a < 1.22$ MeV with more than two orders of magnitude lower integrated cross-section, $\Sigma_i < 1\times10^{-3}$ mb keV; $E_a = 1.22$ MeV, $\Sigma_i = [8.5(+4, -3)]\times10^{-2}$ mb keV; $E_a = 1.88$ MeV, $\Sigma_i = [1.6(+0.8, -0.5)]\times10^{-1}$ mb keV; and $E_a = 2.66$ MeV, $\Sigma_i = 3.8(+2, -1)$ mb keV.

In 1959 Ikeda and Yoshihara [59Ik] (in Japanese) investigated the $^{87}Sr(\gamma, \gamma')^{87}Sr^m$ reaction by the use of a 370 TBq (10 kCi) ^{60}Co γ-ray source and estimated the overall cross-section of the (γ, γ') reaction to be 22 nb. It was suggested that the lowest activation level of ^{87}Sr lies below 1.33 MeV. In his next paper Yoshihara [60Yo] determined the integral cross-section to be 380 ± 80 mb eV, from a relative measurement. He did not calculate the photon flux to excite the nucleus, because the excited level of ^{87}Sr was not clear; thus he used the total photon flux impinging on the target instead of $N_\gamma(E_{AL})$. It should be noted that the value should be doubled because the photoexcitation process in this nuclide can take place only with the higher energy γ-rays from ^{60}Co.

In 1963, Veres also irradiated the ^{87}Sr nucleus, [63Ve2], independent of the previous result of Ikeda and Yoshihara (although he had no information from the experiments published in Japanese). Veres estimated the threshold energy of the (γ, γ') reaction to be about 1.1 MeV for the first activation level. Under the given experimental conditions the calculated number of gamma quanta (Compton-scattered into a 1 eV energy interval per number of gamma quanta from the source) was found to be 1.7×10^{-7} eV^{-1}. Given this value and taking into consideration the fact that the activation level width is much smaller than 1 eV, the integrated cross-section can be computed from the experiment.

The experimental cross-section is 8.5 ± 2.6 nb and the calculated integrated cross-section was found to be 50 ±15 mb eV. From the more precise data [96Fi] the 1.22 MeV activation level corresponds to the 1.2284 MeV level. Taking this into consideration the re-evaluated results of [63Ve2] are: $\sigma_i = 30 \pm 9$ mb eV, and $\tau = 1 \pm 0.3$ ps. This latter lifetime of the 1.2284 MeV level is in agreement with the value of the lifetime data given in the literature [96Fi].

Zaparov et al. [73Za] reported 61.7 mb eV for the integrated cross-section at the activation level of 1.22 MeV.

Watanabe and Mukoyama [82Wa] also excited the ^{87}Sr nucleus with a ^{60}Co radiation source, but they measured larger values than the results of Booth and Brownson and of Veres. Their published data of 420 ± 60 or 560 ± 80 mb eV for the integrated cross-section depend on the assumed values of the respective partial level width $g\Gamma_0 = 1\times10^{-3}$ or $g\Gamma_0 = 1\times10^{-4}$ eV. The results of Watanabe and Mukoyama are based on many assumptions about the nuclear parameters of excited levels, such as energy, total and partial level width and branching ratio of the transition to the isomeric level. They concluded that to obtain more accurate values for the integral cross- sections, further experimental and theoretical studies on these nuclear parameters would be desirable. They also mentioned that it would be interesting to perform a similar experimental study of the photoexcitation process by means of radioactive sources other than ^{60}Co. It can be seen (from *Table 15*) that such experiments were performed by ^{46}Sc, ^{24}Na radioactive sources [73Ve2], and by a ^{116}In^{m} reactor loop [67Ab]. Veres et al. made a general remark from the experiment carried out with the ^{46}Sc radiation source: namely, they established that ^{87}Sr possesses an activation level below 1.12 MeV. Most probably it is the level at 873.3 keV. Knowing this energy and the quantum numbers the lifetime of the activation level at 873.3 keV was determined; it is 0.9±0.27 ps in agreement with the value of the lifetime data given in the literature [96Fi].

Table 15. The excitation cross-sections and other data of the ^{87}Sr nucleus

Irradiation	Excitation E (MeV)	σ_{eff} (nb)	σ_{int} (mb eV)	Σ_{int} (μb keV) or mbarn keV	$\gamma\Gamma_0\Gamma_m/\Gamma$ (MeV)	Lifetime τ (ns or ps)	Refs
3 MeV Van de	<1.22	<0.03		$<1\times10^{-6}$	$<3\times10^{-4}$		[67Bo]
Graaff having	1.22	1 (+0.6, –0.3)		85 (+40, –30)	3.3×10^{-2}		
an E_{res}±10 keV	1.88	1.4 (+0.7, –0.5)		160 (+80, –50)	0.15		
and 150 μA	2.66	4.1 (+0.2, –0.5		[3.8(+2, –1]$\times10^{3}$			
^{60}Co, 740 TBq	<1.33	22					[59Ik]
	1.22		300±80				[60Yo]
^{60}Co, 48 TBq	1.1	8.5	50±15 b)				[63Ve2]
^{116}In^{m} reactor loop	<2.112	57					[66Ab]
^{46}Sc, 16 TBq	0.873	3.2				1.76 ps	[73Ve2]
^{60}Co, 7.8 PBq	1.22		61.7				[73Za]
^{60}Co, 18.5 TBq	1.229		420±60 a)				[82Wa]
^{60}Co, 112 TBq	1.229		47±7				[86Kr1]
Four LINAC 0.5–11	4			3900±130			[91Ca2]
MeV, BS	6			8700±180			
4–15 MeV, BS	8			0.52±0.1*			[93Ma1]
	10			1.6±0.14*			
9–30 MeV, BS				2.9*			[93La1]
4 MeV, LINAC	4	1330±150	8.8±1.3 μb MeV				[96La]

a) calculated from the isomer activity data; b) 420±60 if $g\Gamma_0 = 1\times10^{-3}$ eV, and 560±80 if $g\Gamma_0 = 1\times10^{-4}$ eV

The partial level diagram *(Fig. 35)* shows only levels and transitions below 1.77 MeV. Data are taken from the experiments referred to in *Table 15* and from [78Le], [96Fi]. For example, the A2 level, which can be excited by photons and cascade down to the isomeric state, is known to be at 1.22 MeV [67Bo]. The next higher activation levels (1.88 MeV and 2.66MeV) are omitted

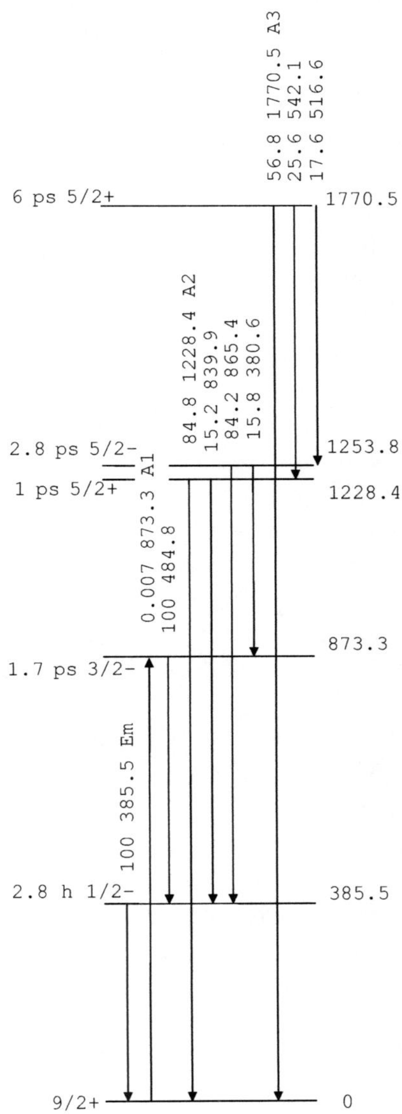

Fig. 35. A partial γ-excitation and decay scheme of ^{87}Sr

from the diagram since there is no satisfactory information about the quantum numbers, branching ratio and transition probabilities. The excitation of the A1 activation level (dashed line) has a very little probability.

Krcmar et al. [86Kr1] studied the contribution of the non-resonant process to the excitation of isomeric level of ^{87}Sr. They applied the best χ^2 method for the comparison of the resonant and non-resonant cross-section values. Their results indicated that, in many experimental arrangements, non-resonant processes make significant contributions to the photoactivation of isomeric levels. Bikit et al. [87Bi], in order to make a more direct comparison between their measurements and that of [86Kr1], used essentially the same technique in the case of ^{111}Cd and they questioned the necessity of introducing any "non-resonant" photoexcitation process.

Carroll et al. [91Ca2] excited the isomers of 19 nuclides. Four accelerators were used as *bremsstrahlung* sources and endpoint energies covered the range from 0.5 to 11 MeV. They used strontium as a benchmark for normalising the isomer activity measurements performed by the four accelerators. The datum at 1.2 MeV represents an upper limit since no fluorescence photons were observed above the background level. This indicates some discrepancy with the results of Booth and Brownson and those of Veres et al. [73Ve2] because they measured isomeric activity above the background by a ^{46}Sc radiation source. It has two γ-lines with energies of 1.12 MeV (100%) and of 0.89 MeV (100%), respectively, and the experiment indicates the existence of an activation level below 1.12 MeV. Carroll et al. [91Ca2] drew some conclusions from the measurement data of the 1.2–6 MeV energy range. One conclusion is that there is no evidence to support the participation of any non-resonant process. This type of mechanism, if present, would be heavily dependent upon the density of nuclear states, which rises sharply at energies approaching E_n. The threshold energy E_n for the (γ, n) reaction is 8.428 MeV. The slow increase in the excitation function above 6 MeV, relative to the change seen below 4 MeV, preludes this as the dominant means of photoexcitation.

Lakosi et al. [95La2] excited the ^{87}Sr isomer by (γ, γ') reaction using *bremsstrahlung* from a 0.9 mm Pt converter placed against the electron beam of a 4 MeV, 25 μA linear accelerator and determined 10 μb MeV as the value for the integrated cross-section.

The results in *Table 15* show a large variety in the magnitude of the cross-sections and demonstrate that the isomer activation is a multiparametric process. The large differences occurring in the magnitude of the cross-section cannot be explained on the basis of a single quantity (e.g., of deformation parameter [92Co]), or spin difference [93Ma1], but rather as a result of an interplay among various factors in a large ensemble.

6.5. ^{89}Y

The natural abundance of the ^{89}Y isomer is 100% and it has an isomeric state at 909.2 keV($9/2^-$), with a half-life of $t_{1/2}$=15.7 seconds. In the literature the examples of only a few experiments can be found, since the nucleus has an extremely low-cross section due to the low level densities of this half-magic nucleus and the high-lying isomeric level.

In 1963 Veres [63Ve2] carried out an experiment irradiating 10 gram yttrium-oxide with a 100 TBq ^{60}Co γ-source. The overall cross-section was <0.4 nb for the energy range of 1.17–1.33 MeV. Zaparov [73Za] determined an integrated cross-section ≤ 0.4 mb eV value using a ^{60}Co source. However, above the background the measured isomer activities were so small that the results were not sufficient to prove the existence of an activation level in the range of 1.17–1.33 MeV.

Abrams and Pelekis [67Ab] irradiated an yttrium target with a $^{116}In^{m}$ source ($E_\gamma = 2112$ keV) produced in a neutron reactor loop, and they measured no isomer activity.

Kaminishi and Kojima [63Ka] produced nuclear isomers of stable nuclides with *bremsstrahlung* from a 6 MeV linear electron accelerator. They published a 9×10^5 count per minute at a beam energy of 5 MeV.

Alston et al. [68Al] studied the resonance fluorescence and carried out absorption measurements, establishing transition strengths in odd-mass nuclei. Resonance fluorescence was seen from a level at 1.51±0.01 MeV. It is assumed that this is the level found at 1.49 MeV and assigned as $3/2^-$ and reported by Stautberg et al. [67St]. The magnitude and sign of the mixing ratio δ are determined from anisotropy in the angular distribution. The measured anisotropy – W(150%) / W(90%) = 1.52±0.12 – is compatible with four choices of δ: $\delta = +0.69\pm0.04$, $\delta = 6\pm1$, $\delta = -0.15\pm0.05$ and $\delta = -1.4\pm0.2$. The electron scattering experiment by Peterson and Alster [68Pe] yielded $B(E2\downarrow) = 60\pm25\ e^2 fm^4$ for the 1.5 MeV transition. Combining this value and the partial level width (52±1.6 meV) given from the absorption measurement of Alston et al., one obtains a mixing ratio of -0.13 ± 0.03. This is consistent with the -0.15 ± 0.005 value measured by Alston et al.

Breban et al. [79Br] also measured the minimum detectable amount of 18 isomers, among them that of yttrium. They determined a minimum concentration of ^{89}Y as 7 microgram/gram.

Carroll et al. [91Ca2] excited the isomeric states of 19 stable nuclides, using four accelerators as *bremsstrahlung* sources with endpoint energies ranging from 0.5 to 11 MeV. The integrated cross-sections of ^{89}Y were found to be 90±50 μb keV at 4 MeV and 2600±80 μb keV at 6 MeV energy.

Huber et al. [93Hu] investigated the $^{89}Y^m$ excitation function for *bremsstrahlung* with endpoint energies E_0 = 2–5 MeV. Up to 2.875 MeV no isomer activity was detected, towards higher energies, breaks of excitation function were clearly visible just below 3 and around 4 MeV. The integrated cross-sections were 0.8 μb MeV and 1.2 μb MeV at 2.9 and 4.0 MeV, respectively.

Lakosi et al. [95La2] measured the excitation of $^{89}Y^m$ under the same circumstances as ^{87}Sr and of the other 22 nuclei examined. The effective cross-section for a flux >1 MeV was 30 nb and the integrated cross-section was 0.2 μb MeV at 4 MeV endpoint energy.

The statements about the large variety of cross-section values are analogous to those mentioned in the case of ^{87}Sr.

6.6. ^{90}Zr

The zirconium element has 5 stable nuclei and among them the only ^{90}Zr nucleus (with a natural abundance of 51.43%) has an isomeric state of 2.319 MeV (5^-), and the half-life of the metastable level is 0.8 sec. This isotope is an even-even nucleus with 40 protons and 50 neutrons.

Kaminishi and Kojima [63Ka] irradiated targets of even-even nuclides such as ^{90}Zr, $^{178,180}Hf$, ^{190}Os, and ^{204}Pb by 5 MeV *bremsstrahlung* of a linear accelerator and produced a 4–6 MeV electron beam; however, no isomer activity could be observed in their experiment. Breban et al. [79Br] also investigated the isomer activation of $^{90}Zr^m$ by *bremsstrahlung* of an electron accelerator of 6–8 MeV and they observed no isomer activity, either. Ramchandran and McIntyre [69Ra] studied the spins and the level-widths of excited levels in the 5–9 MeV energy region. Neutron capture γ-rays have been used to excite one nuclear level in ^{90}Zr, Cd, Sn, Hg, ^{205}Tl, ^{204}Pb, ^{209}Bi, respectively. Angular distribution measurements of scattered γ-rays showed that the transitions were all consistent with dipole radiation. For various reasons the data could also be reproduced in most cases by assuming a mixture of dipole and quadruple radiation. They made three further measurements for the ^{90}Zr excitation:

(i) an absolute differential cross-section measurement on the resonance scattering;
(ii) a resonance measurement, and
(iii) a differential cross-section measurement at liquid nitrogen temperature.

The resonance-scattering experiment is a function of the three parameters Γ (the total line-width of the excited state), Γ_0 (the partial line-width associated

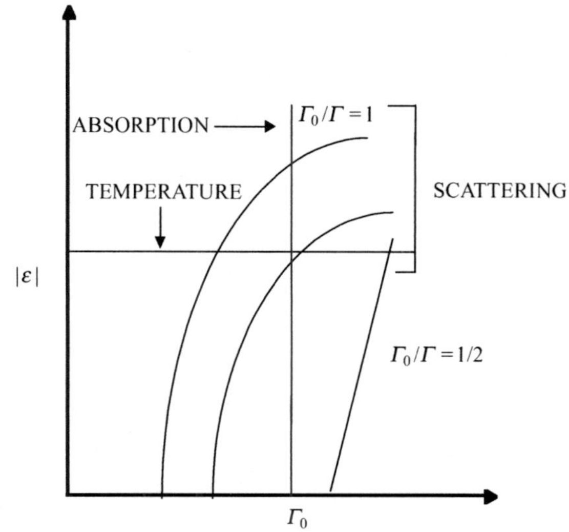

Fig. 36. Plot of the allowed values of Γ_0 and $|\varepsilon|$ for the three different types of experiment: absorption, scattering and temperature variation

with the γ-ray transition to the ground state of the nucleus), and ε (the energy displacement between the neutron-capture γ-ray energy and the resonance energy).

The plot of the allowed values of Γ_0 and $|\varepsilon|$ is shown in *Fig. 36.* The determined data [Ra69] for the 8.496 MeV level of ^{90}Zr are: $\Gamma_0/\Gamma = 0.8\pm0.2$; Γ_0 (eV) = 1.68±0.02; and ε (eV) = 5.60±0.15. Most comprehensive studies have been carried out by Metzger [71Me], [72Me1], [74Me] for the resonant scattering. He observed resonant scattering of *bremsstrahlung* for the 2.186 MeV (the first 2^+ level), 3.308 (2^+), 4.580 ($1^{(-)}$), and 5.504 MeV ($1^{(-)}$) levels. The main results of the Γ_0^2/Γ, Γ_0/Γ, and Γ_0 values in the resonance-fluorescence experiments for the above-mentioned levels (as well as some others) are sum-

Table. 16. Results of resonance-fluorescence experiments with ^{90}Zr

Level E (keV)	I^π	Γ_0^2/Γ (meV)	Γ_0/Γ	Γ_0 (meV)	Γ_0 (expt.)/Γ_0 (s.p.)
2186	2^+	4.89±0.28	1.00	4.89±0.28	5.1
3308±1	2^+	3.45±0.60	0.75±0.02	4.6±0.8	0.6
3842±1	2^+	24.0±3.4	0.86±0.02	28±4	1.7
4580±2	$1^{(-)}$	24±4	0.5±0.1	48±13	3.7×10^{-4}
5504±2	$1^{(-)}$	48±17	0.6±0.1	80±32	3.5×10^{-4}
4120	2^+	0.3±0.5			
4230	2^+	0.5±0.9			
4680	2^+	<10			

marised in *Table 16.* It is possible that the $B(E2)$ values given in these experiments contained a contribution from ^{91}Zr. Moreover, very important information has been obtained for other parameters of some levels of ^{90}Zr. However, in these works there is no information about the cascade transition probability to the isomeric state.

Veres et al. [73Ve2] carried out an experiment using the 2.75 MeV γ-ray from ^{24}Na radiation sources of 100 TBq intensity for the irradiation of a zirconium target. Because of the 0.8 second half-life of the isomeric state of ^{90}Zr, irradiation and measurement were carried out in successive cycles of 30 seconds and 2 seconds, respectively. The samples irradiated with the γ-source could be taken in 1 second by pneumatic tube to the measuring detector. No measurable isomeric activity was obtained for the zirconium target under investigation. Therefore the only useful information furnished by the experiment is that the upper limit of the activation cross-section per primary γ-quantum could be estimated as <0.02 μb for $^{90}Zr^{m}$.

There are no sufficient experimental data about the cross-section values, activation level energy, quantum numbers of the excited levels and other parameters such that an evaluation can be made of the role of zirconium in graser developments.

6.7. ^{93}Nb

The ^{93}Nb is the only stable isotope of the element niobium. The spin and parity of the ground state is $9/2^-$. The half-life of the metastable level of 32.82 keV($1/2^-$) is 16.1 y. In the γ-decay scheme from 1 MeV to 2.7 MeV [96Fi], the excited levels of the 959 keV, 979 keV, 1336 keV, 1491 keV, and 2180 keV energy have $13/2^+$, $11/3^-$, $17/2^-$, $15/2^+$, and $19/2^-$ spin/parity values, respectively. Therefore our knowledge about the cross-section of the $^{93}Nb(\gamma, \gamma')^{93}Nb^{m}$ reaction above several MeV is rather scanty.

In 1945 Wiedenbeck [45Wi1] bombarded ^{93}Cb for eighteen hours with X-rays produced by a beam of 150 μA 2.6 MeV electrons, stopped on a gold target. A small sample of columbium showed an initial activity of 2000 counts per hour when the sample was placed to a thin windowed (1.5 mil Al) counter. This activity was found to decay with a half-life of approximately 42 d. Later, the half-life of the $^{93}Nb^{m}$ ($^{93}Cb^{m}$) was adopted as 16.1 y.

In the energy range below 1 MeV, several levels were found: 949 keV – $13/2^+$, 810.4 keV – $5/2^-$, 808.6 keV – $5/2^-$, and 743.9 keV – $7/2^-$. These could be excited with $E2$ or $M2$ transitions. Veres et al. [69Ve] attempted to excite the isomeric state of the ^{93}Nb. Four niobium targets – each with a diameter of 12 mm, a thickness of 0.0566 mm and a weight of about 120 mg – were irradiated

Table 17. γ-line $E\gamma$, level E_l energies, intensities and spins J^π of ^{93}Nb [79Br]

$E\gamma(\pm3)$ keV	Intensity	E_l (±3) keV	Spins J^π
5516	280± 5	949	$13/2^+$
5486	2± 2	979	9/2,11/2,13/2
5384	47± 2	1081	9/2,13/2
5168	30± 2	1297	9/2,13/2
4979	12± 3	1486	9/2,11/2,13/2
4971	26± 1	1494	9/2,11/2,13/2
4783	31± 1	1682	9/2,11/2,13/2
4514	32± 3	1951	11/2
4095	80±25	2370	9/2,13/2
3626	80±25	2839	11/2
6465	100± 2	6465	$11/2^+$

with a 1.13×10^{13} γcm^{-2} s^{-1} intensity of a ^{60}Co source. The irradiation lasted for 134 d, and the activity was measured with a set-up employing a scintillation detector with an approximately 2π solid angle. The detector was calibrated with the 20.5 keV characteristic X-rays of $^{103}Rh^m$. Non-irradiated inactive niobium plates were used to determine the background. The measurements were repeated many times but very little (0.08±0.03 decay per second) isomeric activity was observed above the background. After re-evaluating the experimental data of the measurements from 1969 with new quantum numbers and a γ-decay scheme, the lifetime of the 810 keV level was determined as 40±20 ps.

Breban et al. [79Br] studied the energy levels of ^{93}Nb using photoexcitation. The photon source was produced by neutron capture reaction. The intensity of the strong lines of the $V(n, \gamma)$ and $Ti(n, \gamma)$ sources was $\approx10^6$ monoenergetic photon/cm^2 sec. The excitation and decay of the 6465 keV level of ^{93}Nb (showing level energies and relative intensities normalised so that the intensity of the ground-state transition is taken to be 100) are given in *Table 17.* However, in this experiment no information could be found about the transitions to the isomeric state.

Sáfár and Lakosi [93Sá1] calculated the cross-section of the $^{93}Nb(\gamma, \gamma')$ $^{93}Nb^m$ reaction in the 5 to 30 MeV γ-energy region, in the framework of a spin-dependent gamma-ray cascade model. The excited states of ^{93}Nb were described by discrete levels up to a threshold energy of 1.5 MeV and by a level continuum formalism above that. The calculated cross-section curve for the reaction shows a pronounced peak at neutron separation energy with an integral of 3.1 μb MeV up to 12 MeV. The order of magnitude of the smaller contribution (multiplied by 1000) of the pre-equilibrium γ-ray emission (having a maximum at ~18 MeV) can also seen in *Fig. 37.* The calculated excitation function makes it possible to

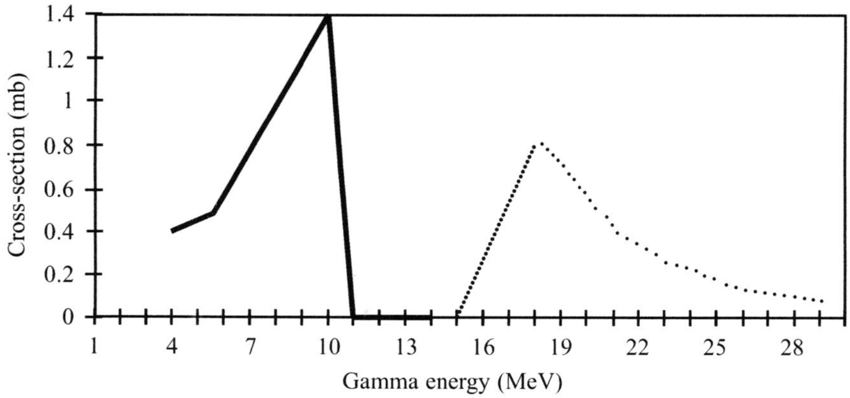

Fig. 37. Calculated excitation function of the isomeric state of the ^{93}Nb in the 5–30 MeV γ-energy region. The contribution of the pre-equilibrium γ-ray emission is multiplied by 1000 (the dashed line)

get an estimate for the effect of capture γ-rays on neutron-influence monitoring. For the case of a 30 MeV electron beam and 2π neutron detection geometry, the ratio $R_{\gamma/n}$ proved to be $\approx$13.

In another paper Lakosi and Sáfár [93La1], realising the situation that the generally accepted long-standing picture of a two-humped excitation function for photo-production of isomers had been disproved, made semi-empirical estimates for saturated values of the integrated cross-section. They summarised the results for 31 stable nuclei having isomers with long lives. For the ^{93}Nb nucleus, given the lack of experimental data for the integrated cross-section, the estimation gives 3.9 μb MeV. This agrees reasonably well with the 3.1 μb MeV value of the more advanced calculation [93Sá1].

6.8. ^{99}Tc

The actual discovery of the element 43 had to await the development of nuclear physics and the modern cyclotron. Perrier and Segré [37Pe] used the Berkeley cyclotron to bombard a molybdenum sample with deuterons and subsequently detected γ-radiation which they ascribed to the element 43.

The discoverers named it technetium, because it was the first element that had been produced artificially. Little is known about the nuclear properties of technetium, mainly due to its absence from terrestrial nature; this is because the half-lives of its longest isotopes are: $t_{1/2} = 2.6\times10^6$ y for ^{97}Tc, $t_{1/2} = 4.2\times10^6$ y for ^{98}Tc, and $t_{1/2} = 2.1\times10^5$ y for ^{99}Tc. The latter, in the ground state, has a $9/2^+$ spin/parity and its isomeric state has a $1/2^-$ spin/parity. The energy of the isomeric state is 142.7 keV, and the half-life of the isomer is 6.01 hours.

6.8.1. Experimental and calculated results

In 1978 Veres and co-workers [unpublished experiment] attempted to excite the isomeric state of ^{99}Tc with the γ-rays of a ^{60}Co source. About 0.4 gram (780 GBq; 20 mCi) of ^{99}Tc was irradiated with a 100 TBq source. Because the background was very high as a result of the intense *bremsstrahlung* of the target from the β^- decay of ^{99}Tc ($\beta_{max} = 292$ keV), isomer activity above the background could not be observed.

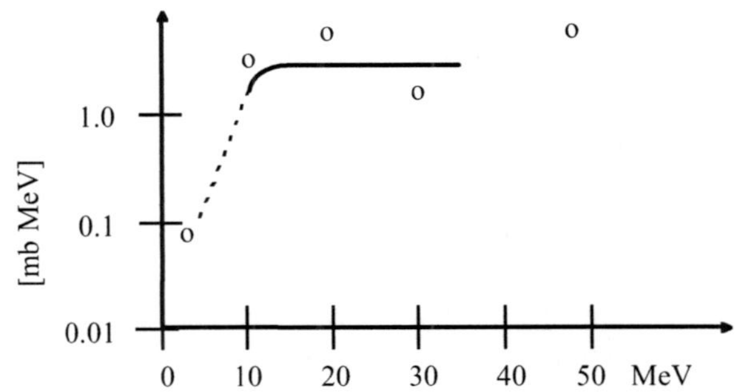

Fig. 38. Measured (o) and calculated (solid line) integrated cross-section for the ^{99}Tc(γ, γ')^{99}Tc^m reaction in the function of endpoint energy

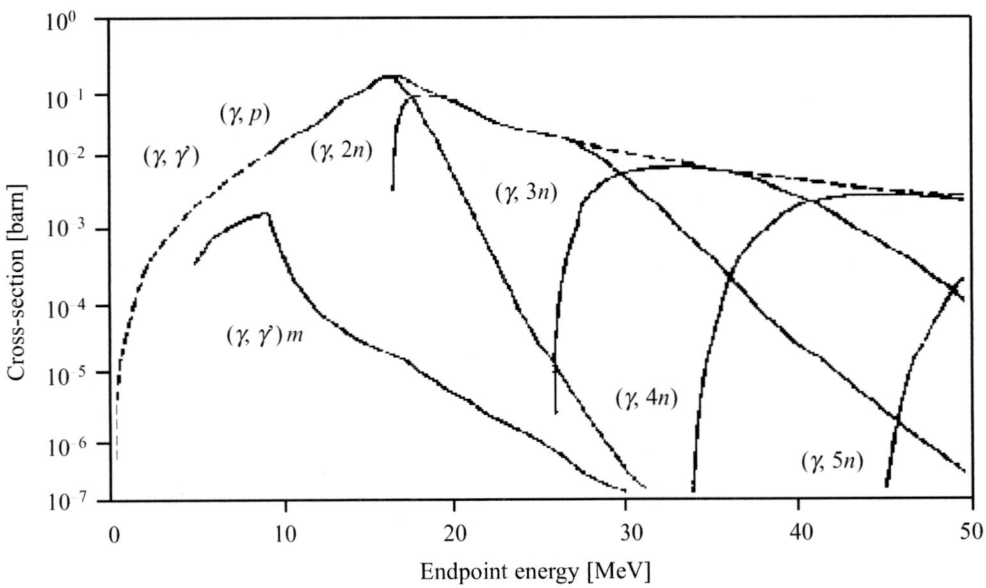

Fig. 39. Calculated photoneutron and (γ, γ')^m cross-sections for ^{99}Tc. The total photoabsorption cross-section, based on the Lorentz parameters, is drawn with a dashed line

The level structure of ^{99}Tc offers an outstanding possibility to study the induced depopulation, by a very low energy (2.2 keV) transition, of the long-lived (6 h) ^{99}Tc^{m} in an intense radiation field.

The stimulated emission of 2.2 keV photons upon being irradiated by synchrotron radiation was considered. An estimate for the critical flux density led to about ten orders of magnitude higher than the available flux values. Experimental and calculated integral cross-section values were determined from the saturation activities obtained at 15, 20, 30 and 50 MeV energy, respectively (Sendai) [91Se] and at 4 MeV endpoint energy (Budapest), as plotted in *Fig. 38.*

The integral cross-section, as seen in *Fig. 38,* turns to saturation at about the photoneutron threshold, 9 MeV, and then remains practically constant, even around giant resonance.

The calculated photoneutron and (γ, γ') cross-sections have also been plotted for individual reactions from the threshold energies up to 50 MeV (see *Fig. 39*). The calculation was performed in the framework of a cascade-evaporation model. The nuclear photoabsorption cross-section was taken as a standard Lorentzian, multiplied by the probability of evaporation of one or more neutrons.

A more favourable mechanism for induced depopulation may be the stimulated emission of a virtual photon, leading to the ejection of an orbital electron instead of a real photon (utilising the exceptionally high conversion coefficient). Another possibility for nuclear de-excitation may be the electron-bridge mechanism. It refers to a third-order correction to internal conversion, in which emission of a real photon and the simultaneous ejection of an orbital electron are considered; this is a consequence of the exchange of a virtual photon between the electron and the nucleus [93La2].

The partial level scheme *(Fig. 40)* shows only levels and transitions up to 1072.2 keV. Data are taken from the references [78Le], [96Fi]. For example, the A1 and A2 levels, which can be excited by photons and can cascade down to the isomeric state, are known to be at 761.7 keV and 1072.2 keV, respectively.

In 1993 Lakosi et al. [93La2], using *bremsstrahlung* up to 50 MeV, determined the (γ, *xn*) and (γ, γ') reaction cross-sections of ^{99}Tc for the first time by means of γ-spectrometry. Irradiation was performed by the linear electron accelerator of the Laboratory of Nuclear Science, Tohoku University, Sendai, and the 4 MeV accelerator of the Institute for Isotopes, Budapest. The calculation was also carried out in the framework of a neutron evaporation and a statistical gamma-ray cascade model. The 140 keV line in the γ-spectrum comes from the decay of the 6 hour half-life isomer, which is produced by the (γ, γ') reaction; the threshold of the latter was not known exactly. It should, however, lie below 4 MeV, because photoexcitation of ^{99}Tc^{m} was observed at irradiation by 4 MeV endpoint energy *bremsstrahlung*.

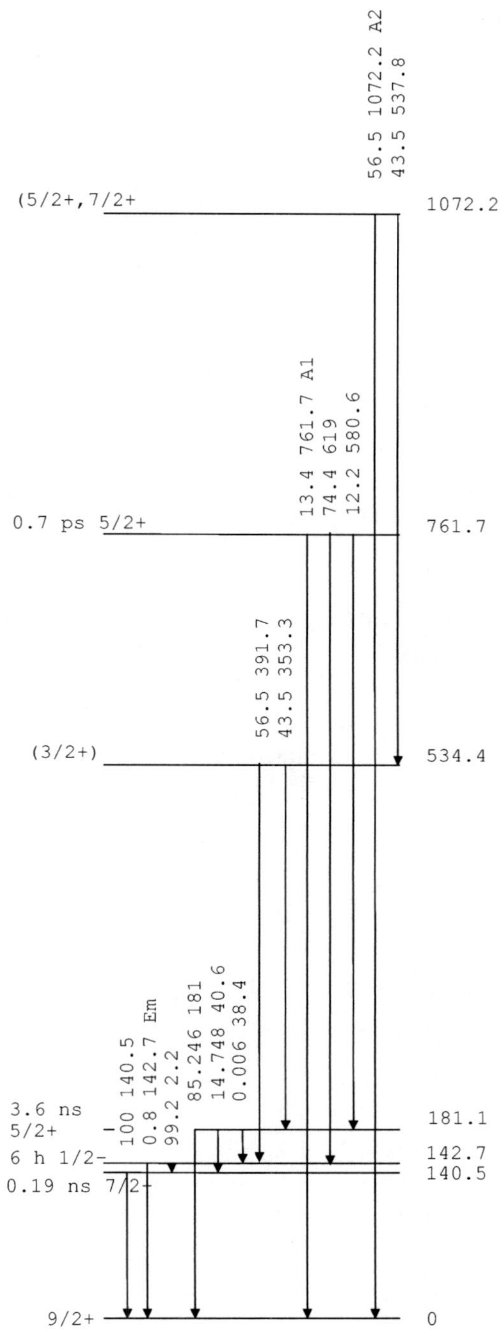

Fig. 40. Partial γ-decay scheme of ^{99}Tc

Above photoneutron threshold the neutron emission channel was assumed to be the only channel open for de-excitation following photoabsorption. The pre-equilibrium contribution was negligible. For more details, see [89Se1], [94Se].

A series of 25 isomers were excited by a (γ, γ') reaction, using *bremsstrahlung* from a 0.9 mm Pt converter placed against the electron beam of a 4 MeV, 25 μA linear accelerator [95La2]. Among them the integral cross-section of ^{99}Tc^{m} was determined by 70 μb MeV for the endpoint energy of *bremsstrahlung* (see *Fig. 38*). In some practical cases [88Yo], however, the cross-section can be interesting in the lower energy region as well. According to the measurements of Ikeda et al. [88Ik], the reactor-neutron-spectrum-averaged cross-section of the ^{99}Tc(n, n')^{99}Tc^{m} reaction turned out to be 240 μb.

6.8.2. Implication of photonuclear reactions of ^{99}Tc in nucleosynthesis

Under stellar conditions, exotic nuclear phenomena may occur, like the β-decay of nuclear excited states in ^{99}Tc; these should be considered in the theory of nucleosynthesis. There are a number of possible ways in which the technetium isotopes having long lives could be produced in a stellar interior. The most acceptable one is the s-process in nucleosynthesis. A part of the s-process path is shown in *Fig. 41*. The branch at ^{99}Tc is due to competition between neutron capture and β-decay.

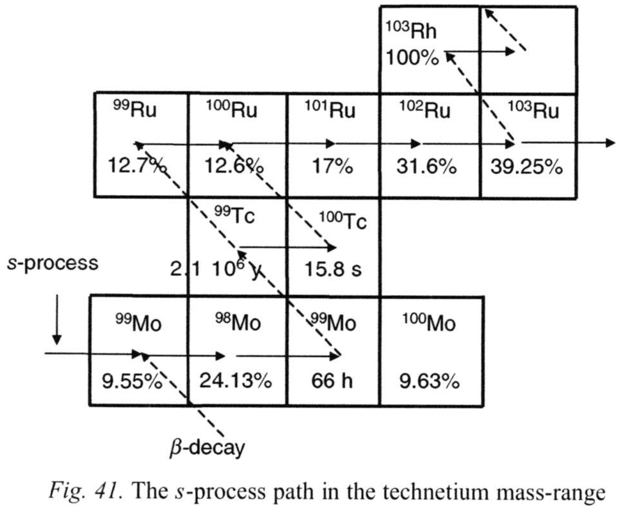

Fig. 41. The s-process path in the technetium mass-range

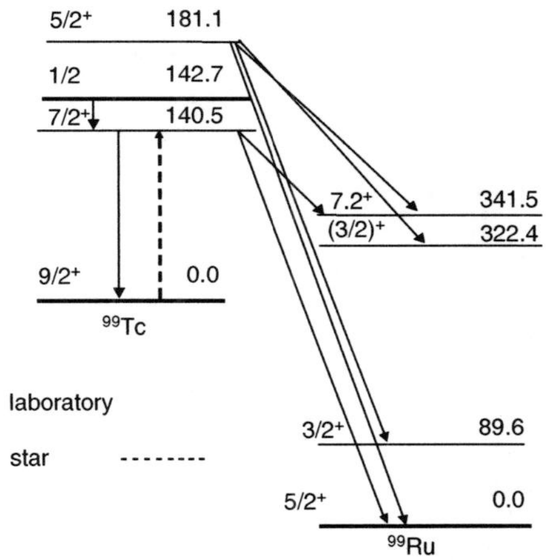

Fig. 42. Partial decay scheme of ^{99}Tc ground and isomeric states. Energies are in keV

Two other mechanisms have been proposed as well: (a) photo-beta decay from ^{97}Mo [59Ca], and (b) the photoneutron emission from ^{100}Mo [74Ed], [77Ch]. For the latter mechanism the calculated (γ, xn) reaction cross-section can be seen in *Fig. 39.* A detail of the level scheme in *Fig. 42* indicates that the long half-life of the ^{99}Tc ground state is due to its decay to ^{99}Ru by a second-forbidden transition; on the other hand, the allowed β-decays from excited levels have much shorter partial half-lives, being in the order of days. Thus the laboratory half-life can be drastically reduced at high temperatures in a stellar interior because of the enhanced β-decay from thermally populated photoexcited states. The relevant β-decays from excited levels are not known experimentally. Detailed shell model calculations taking into consideration both continuum- and bound-state β-decays show that the effective half-life starts to decrease at about 10^8 K and reduces to about 10 y [87Ta] at an average *s*-process temperature of $(3.8\pm0.5)\times10^8$ K [90Kä]. Jung et al. [93Ju] pointed out the relevance of bound-state β-decay in nucleosynthesis.

According to the observations of certain stars, in spite of the enormously enhanced decay rate, a substantial amount of ^{99}Tc survives the hot environments of the *s*-process. The time scale of the order of years indicated above is comparable or shorter than the duration of the thermal-pulse *s*-process environment, which is considered to be 10–100 years. This dilemma is called the "^{99}Tc problem".

The solution of the problem is provided by numerical network calculations which take into account the competition between neutron capture and β-decay. According to these calculations the neutron production is even more enhanced at such temperatures and almost compensates for the depletion [86Ma]. In other words, the equilibrium abundance of ^{99}Tc is very close to the value it would have been if it were stable.

In addition, according to Mathews et al. [86Ma], the observed ^{99}Tc abundance, in comparison with that of Nb and Mo, could be used as an indicator of the life-time of stellar mixing.

6.8.3. Possibilities of the application of ^{99}Tc^{m} in graser developments

In section 5 a rough draft was given of the main proposals which rely upon the comprehensive bibliography. When nuclear isomers are bound in solids, the recoil momentum of gamma-ray emission is imparted to the remainder lattice. The graser dilemma is the need to activate, separate, concentrate, and crystallise an appreciable number of isomeric nuclei by conventional radiochemistry in at least a few seconds; it usually requires several hours. Using a long-lived isomer the first problem is the need for line narrowing, and it would be necessary to work with highly radioactive material (due to the ground state decay of ^{99}Tc). It is not yet known whether highly radioactive crystals can be grown and, if so, the extent to which they would retain their structure.

Besides having the ^{99}Tc(γ, γ')^{99}Tc^{m} reaction for a relatively low cross-section for nuclear excitation, the photons are intrinsically inefficient. Therefore, ^{98}Mo(n, γ)^{99}Mo (decay to ^{99}Tc^{m}) or ^{100}Mo(γ, n)^{99}Mo (also decay to ^{99}Tc^{m}) reactions are used to produce ^{99}Tc^{m}. In the (n, γ) reaction the n-capture cross-section is about 130 μb for ^{98}Mo, and as can be seen from *Figs. 16* and *17,* the cross-section of (γ, n) is about 70 μb using a ^{100}Mo target for the irradiation with γ-energy of about 18 MeV. In both cases the ^{99}Mo (half-life: 65.94 hours) decays to the isomeric state of ^{99}Tc with branching of 82.4%.

The idea is that it may stimulate the nucleus from the isomeric state (142.7 keV) to the 181.1 keV excited state with ΔE = 38.4 keV trigger energy. This level has a very short lifetime (3.6 ns), thus a considerable part of the storage energy of the isomeric state could be released by a sudden transition to the ground state *(Fig. 43)*.

It can be deduced from *Fig. 43.* and from section 5 that the stimulation process may be described as $\sigma_{mi} = f\,\sigma_{exci}$, where $f = f_s f_E f_L$, and it depends on the spin, the transition energy, and on the multipolarity – i.e., $f_s \approx (2J_i+1)/(2J_f+1) \approx (2\times 7/2+1)/(2\times 1/2+1) \approx 4$; $f_E \approx [(E_{exc} - E_m)/(E_{exc} - E_0)]^{2L+1} \approx 4.3\times 10^{-4}$; $f_L \approx [(L_1+1)/L_1\,[(2L_1+1)!!]^2]/[(L_2+1)/L[2L_2+1)!!]^2] \approx 2$; and $f \approx 2.4\times 10^{-3}$.

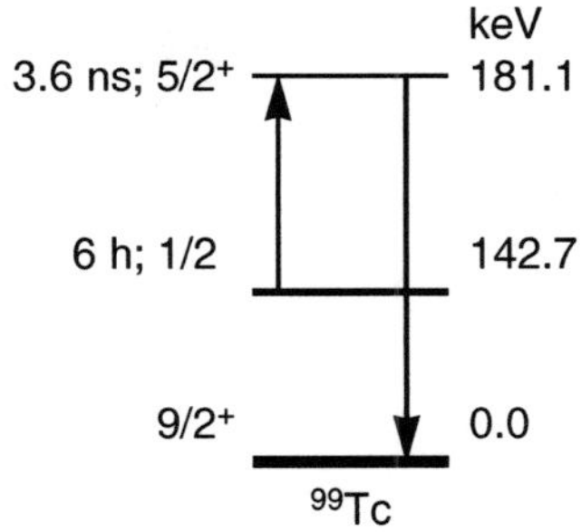

Fig. 43. A simple scheme of inversionless lasing of $^{99}Tc^m$

The only experiment of this type was carried out by Bikit et al. [93Bi]. They investigated the depopulation of $^{99}Tc^m$ by the *bremsstrahlung* of a 15 MeV LINAC. Their idea was also to find out – in a similar way to the above-mentioned process, but with a calculated trigger energy of 3 MeV – whether the decay rate of $^{99}Tc^m$ would be influenced observably by placing the target in a strong radiation field. In their experiment they used four ^{99}Tc targets in the form of sodium pertechnetate, each of them having about 1.5 GBq (40 mCi) activity. For the determination of the irradiation effect the authors compared the activity decrease with non-irradiated (A_0) and irradiated, (A_i) targets. As a measure of the possible effect of irradiation, the relative difference from the reference case was used. The photon fluxes of the irradiation were 1.1×10^7 [cm^{-2} s^{-1} keV^{-1}] and 2.9×10^7 [cm^{-2} s^{-1} keV^{-1}] at distances from the *bremsstrahlung* target of 37 cm and 23 cm, respectively, at 3 MeV energy. The cross-sections they obtained in this way were $\sigma_{37\ \mathrm{cm}} = 2(1) \times 10^{-14}$ [cm^2 keV] and $\sigma_{23\ \mathrm{cm}} = 1.9(3) \times 10^{-14}$ [cm^2 keV].

The authors estimated a depopulation effect of up to 2%; this would be checked independently by some other method.

6.9. ^{103}Rh

^{103}Rh also is the only stable isotope of the rhodium element, and it has an isomer state of 39.756 keV energy and spin/parity of $7/2^+$. Its half-life is 56.12 minutes. In many experiments – because of the very high internal conversion coefficient of the transition to the ground state, $1/2^-$ – the characteristic 20.5 keV X-ray emission of rhodium was used to determine the isomeric activities of $^{103}Rh^m$. The (γ, γ') excitation was studied by Wiedenbeck [45Wi]. He observed activation levels at 1.26 MeV (as a threshold energy), and at 1.64, 2.02, 2.37, 2.71, and 3.05 MeV energies. When the energy of the monokinetic electron beam,

Table 18. The excitation cross-sections and other data of the ^{103}Rh nucleus

Irradiation facility	Excitation E_{exc} (MeV)	σ_{eff} (nb)	σ_{int} (mb eV)	Σ_{int} (mb MeV)	Refs
^{60}Co, 370 TBq	1.26		220±110		[60Yo]
^{60}Co, 48 TBq	1.1; 0.88 [a)]	0.8	4.5±1.5a		[63Ve2]
^{116}In^{m} reactor loop	<2.112	5±1			[66Ab]
^{24}Na, 5 TBq	<2.75	<3			[69Ab]
^{60}Co, 7.8 PBq	1.26		3		[73Za]
^{64}Cu, ~06 TBq	1.26	1.3±0.3 μb [b)]			[83Vi]
Synchr. radiation	0.295c		21±8		[89Yo]
^{60}Co, 2.8 PBq	1.2557	134±43	170±55		[95La1]
50 MeV, LINAC	14–30			24.3 [d)]	[91Sá]
4 MeV, LINAC	4 endpoint	27.2±3.8 μb		0.181±0.028	[96La]

a) re-evaluated data; b) σ_{eff} = 1.3±0.3 μb, σ_{res} = 9.1±2.4 barn, and σ_{nonres} = 0.27±0.07 mb; c) 0.295 or 0.357; d) This is a calculated value, while 22.6±4 mb MeV was obtained for the mean experimental value

which produced the X-rays, reached a higher activation level, the yield curve showed a sharp increase in slope.

^{60}Co γ-rays were used by Ikeda and Yoshihara [56Ik], [59Ik] for the excitation of ^{103}Rh; in a following paper [60Yo] Yoshihara found the integral cross-section to be (220±110) μb eV at the 1.26 MeV energy level.

Veres [63Ve2], independently, also measured the experimental data of ^{103}Rh^{m} and the integral cross-section was found to be (4.5±1.5) μb eV at the 880.5 keV level. Zaparov [73Za] found that the integral cross-section of ^{103}Rh^{m} is 3 μb eV at a 1260 keV energy level.

Vishnevsky et al. [83Vi] observed the isomeric activity of a ^{103}Rh nucleus in positron annihilation. They determined the effective, resonant and non-resonant cross-section values obtained by positron irradiation from a 0.3–0.9 TBq ^{64}Cu as follows: σ_{eff} = 1.3±0.3 μb, σ_{res} = 9.1±2.4 barn, and σ_{nonres} = 0.27±0.07 mb at 1260 keV level energy. Yoshihara et al. [89Yo] studied the possibility of excitation of low-lying ^{103}Rh. The integral cross-section for isomer production by resonance absorption was found to be 21±8 mb eV. No transition from the 295 keV level to the isomeric state has been observed. More accurate information about nuclear properties of 295 and 357 keV levels is necessary.

Sáfár et al. [91Sá] studied the production of ^{103}Rh^{m} by inelastic gamma-scattering in the giant dipole resonance region. The activation method was used to obtain the integral cross-section of the $(\gamma, \gamma')^{m}$ reaction as a function of the beam energy. The *bremsstrahlung* energies for nuclear excitation were 14, 15, 17, 19, 20, and 30 MeV from the ELINAC at Laboratory of Nuclear Science, Tohoku University. Measured integral cross-section of the ^{103}Rh$(\gamma, \gamma')^{103}$Rh^{m}

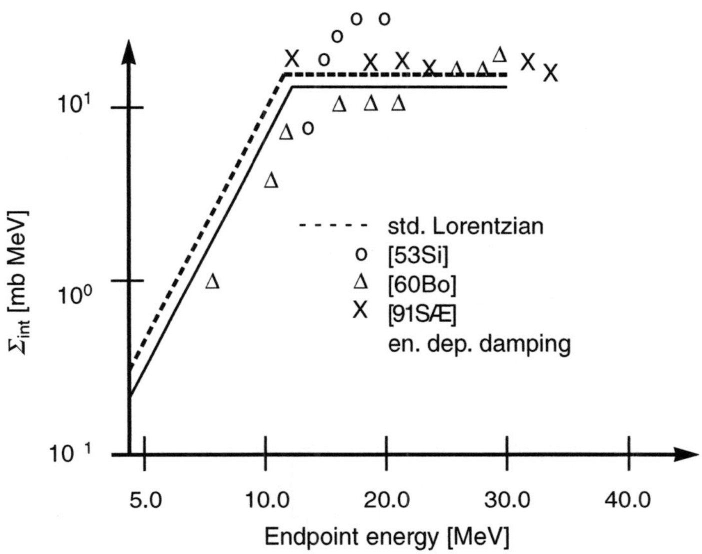

Fig. 44. Energy dependence of experimental (various symbols) and calculated (by Lorentz with energy dependent damping width, solid line; by standard Lorentz, dashed line) integral cross section for $^{103}Rh^{m}$ production [93La3]

reaction in the 10–30 MeV energy range proved to be nearly constant as a function of the endpoint energy [89Se2]. The same holds for the calculated result, as the pre-equilibrium correction to the transition probability provided only a negligible contribution to the integral cross-section. Lakosi et al. [95La1] made a comparison between the calculated and experimental results of the integral cross-section in the 5–50 MeV energy range. The results are tabulated in *Table 18* and plotted in *Fig. 44.*

Consequently, neither the experiments nor the calculations confirmed the existence of a comparable second peak in the photoexcitation function around the giant dipole resonance maximum.

As shown in *Fig. 45,* there are many energy levels of ^{103}Rh in the energy region which are below 1277 keV. In the decay scheme some spins, parities, and lifetimes of these levels are shown which are linked to the isomeric state via activation levels.

Although many practical applications of the isomers have been studied, developments in the field of laser techniques have not as yet appeared in the literature.

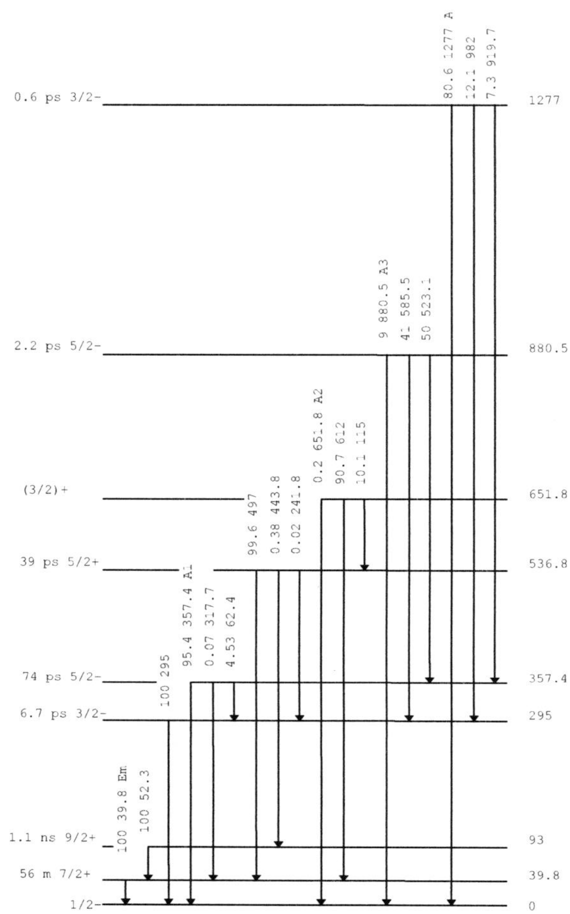

Fig. 45. Partial decay scheme of ^{103}Rh. Energy is shown in keV. Data are taken from the literature [96Fi]

6.10. ^{107}Ag, ^{109}Ag

The element silver has two stable isotopes: ^{107}Ag and ^{109}Ag; their natural abundances are 51.84% and 48.16%, respectively. Both have isomeric states with half-lives of 44.3 and 39.6 seconds, respectively. The energy and spin/parity values of the isomeric states of ^{107}Ag and ^{109}Ag are 93.13 keV($7/2^+$) and 88.04 keV($7/2^+$), respectively.

It had previously been shown by Feldmaier and Collins [41Fe] that the isomeric activity of ^{107,109}Ag can be produced by direct X-ray bombardment of a silver target. Wiedenbeck [45Wi] has found that the excitation function shows

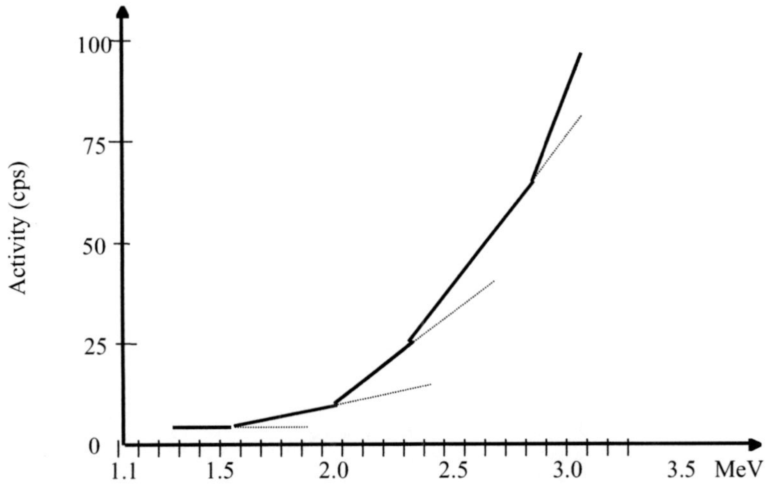

Fig. 46. X-ray excitation of silver. The positions of sudden change in the slope give the position of the energy levels which combine the metastable state

Table 19. The excitation cross-sections and other data of $^{107,\,109}Ag$

Irradiation facility; or isomer activity	Excitation E_{exc} (keV)	σ_{eff} (nb)	σ_{int} (mb eV)	Σ_{int} (μb keV)	g $\Gamma_0\,\Gamma_m/\Gamma$ (meV)	Refs
2 MeV electron accelerator; $^{107}Ag^m$	423±2			0.4±0.3/0.1	2×10^{-5}	[66Ca], [69Bo]
	780±10/5			1±0.7/0.2	2×10^{-4}	
	945±5			3±2/0.5	7×10^{-4}	
	1250±5			30±20/5	1.3×10^{-2}	
	1325±5			90±40/10	4×10^{-2}	
2 MeV electron accelerator; $^{109}Ag^m$	410±2			0.5±0.3/0.1	2.2×10^{-5}	[66Ca], [69Bo]
	680±30			0.6±0.4/0.1	7×10^{-5}	
	1210±10			3.5±2.5/0.5	6×10^{-4}	
	1480±10			85±50/10	3.2×10^{-2}	
	1675±10			130±100/20	9×10^{-2}	
^{60}Co, 740 TBq	1.217	9–12	220±110			[58Ik], [60Yo]
^{60}Co, 48 TBq	1100	8	47			[63Ve2]
^{60}Co, 7.8 PBq			62 [a)]; 93 [b)]			[73Za]
^{64}Cu, 0.3–0.9 TBq	1325, 1210	1.1×10^{5} [c)]				[83Vi]
^{24}Na, 7,48 TBq	<2750	<28; 83				[69Ab], [73Ve2]
4 MeV, LINAC	4 MeV endpoint	4000		34 μb MeV		[95La2]
50 MeV, LINAC	7–9 MeV			21mb MeV		[93La1]

Enriched target: a) ^{107}Ag; b) ^{109}Ag; c) the values for resonant and non-resonant processes are in the text

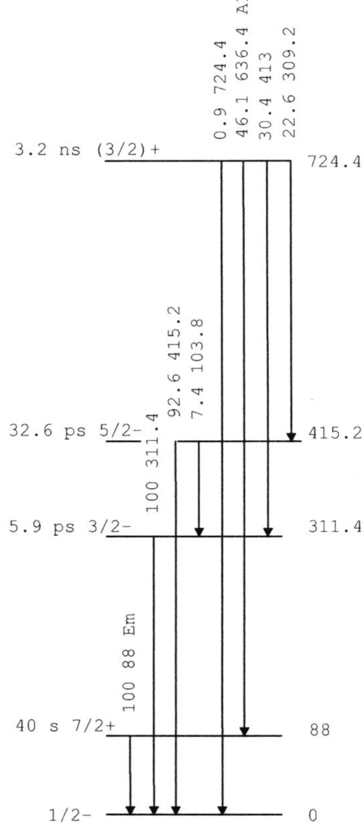

Fig. 47. The partial excitation and decay scheme of ^{109}Ag

sharp increases in its slope. The threshold, which represents the first activation level above the metastable state, is found at 1.18 MeV. Higher activation levels appear at the sharp breaks in the curve: at 1.59, 1.95, 2.32, 2.76, and 3.13 MeV. The curve thus obtained is shown in *Fig. 46.* The half-life was measured to be 40.4±0.2 seconds, representing a mean value for two isomers.

Bizina et al. [63Bi], Cauchois et al. [66Ca], and Boivin et al. [69Bo] also studied $^{107,\ 109}$Ag using *bremsstrahlung* by a 2 MeV linear electron accelerator. The activation cross-sections and the partial level widths were each calculated for the observed levels (data are in *Table 19*).

When radioactive sources are used, γ-quanta of energy suitable for $(\gamma, \gamma')^m$ excitation are produced as a result of Compton-scattering in the source-, intermediate- and target-material. Ikeda and Yoshihara [58Ik], Yoshihara [60Yo], and Veres [63Ve2] also irradiated silver targets by ^{60}Co. The activities of the sources were 1310 and 10000 Ci (~48 and 370 TBq, respectively). The effective cross-

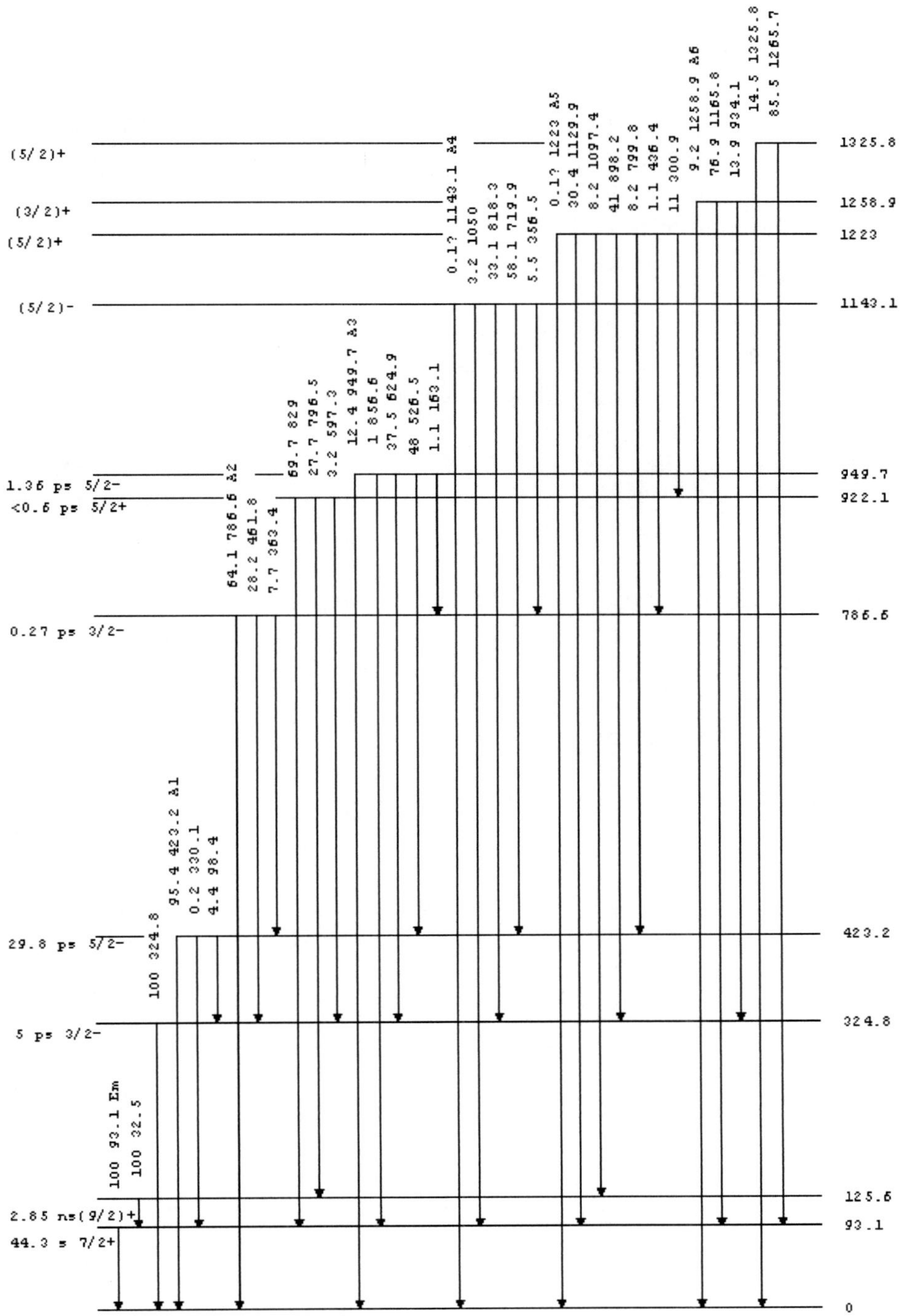

Fig. 48. The partial excitation and decay scheme of ^{107}Ag

sections were measured as 9–12 nb, and 8 nb, respectively. The integral cross-section was calculated for the estimated activation level energy of 1.18 MeV by Yoshihara, and for the estimated activation level energy of 1.1 MeV by Veres. The calculated values were 130–180 mb eV, and 47 mb eV, respectively, representing mean values for the two isomers.

Abrams and Pelekis [69Ab], using a $^{116}In^{m}$ radioactive radiation source produced in a neutron reactor-loop determined a value of 34 mb eV below 2.112 MeV energy.

Vishnevsky et al. [83Vi] excited the isomeric states of $^{107,109}Ag$ by positron annihilation. They determined the effective, resonant and non-resonant excitation cross-sections and compared the results with theory. The results were: $\sigma_{eff} = 110\pm$ 30 mb for $^{107,\,109}Ag$, $\sigma_{res} = 40\pm12$ barn eV for the level of 1325 keV of ^{107}Ag; 40 ± 12 barn eV for the level of 1210 keV of ^{109}Ag; $\sigma_{nonres} = 81\pm24$ mb for the level of 1325 keV of ^{107}Ag; and 81 ± 24 mb for the level of 1210 keV of ^{109}Ag.

Abrams and Lakosi [69Ab], Veres et al. [73Ve2], measured the cross-section of $^{107,\,109}Ag$ with ^{24}Na radioactive radiation sources. They determined the total cross-section values, below 2.75 MeV energy, as <28 and 83 nano-barn, respectively.

Lakosi and Sáfár [93La1] made semi-empirical estimates for isomer excitation above the (γ, n) threshold for 31 stable nuclides having isomers with long lives. They estimated the integral cross-section of ^{107}Ag as 21 mb MeV.

A comparison of the measured integral cross-sections and the energies of the activation levels (with some other data concerning $^{107}Ag^{m}$ and of $^{109}Ag^{m}$) is listed in *Table 19.* In the case of ^{109}Ag, in the low energy range the only activation level is at 724.4 keV, as shown in *Fig. 47.* For ^{107}Ag seven activation levels can be identified *(Fig. 48.),* based on the data of the *Table of Isotopes* 8th Ed. [96Fi]. In the higher energy range the integral cross-section values are orders of magnitude larger than the values that usually characterise (γ, γ') reactions, as is also illustrated in the case of ^{99}Tc (see *Fig. 38*).

6.11. ^{111}Cd and ^{113}Cd

There are eight natural isotopes of cadmium: two (^{111}Cd and ^{113}Cd) with a spin of $1/2^{+}$ in the ground state and a spin of $11/2^{-}$ in the isomeric state. The energies and half-lives of the isomeric states are 392.22 keV; 48.54 m and 263.59 keV; 14.1 y, respectively. The natural abundance is 12.8% for ^{111}Cd and 12.22% for ^{113}Cd.

The photoexcitation of the $^{111}Cd^{m}$ has been studied by many authors: [41Fe], [45Wi], [58Ik], [60Yo], [63Ve1], [65Ch], [66Ca], [67Bo], [69Ab],

[69Bo], [73Ve2], [73Za], [78Ve1], [79Vi], [82Kr], [82Wa], [87Bi], [87Né], [88Né], [91Ca2], [93La1], [93Ma1,2], [91Ne], [93Sá2], [94Né2], [95La2], [96La], [97Ve].

6.11.1. Experimental and calculated data of isomer activation

The position of the activated levels has been found [45Wi] to be similar to the curve shown in *Fig. 46* for silver. The threshold for the ^{111}Cd(γ, γ')^{111}Cd^{m} reaction is 1.25±0.03 MeV and higher activation levels are found at 1.68, 2.08, and 2.56 MeV, respectively. Some of the above-mentioned experiments, in which the cross-section and other data were determined, are listed in *Table 20.*

Table 20. The excitation cross-sections and other data of the ^{111}Cd nucleus

Irradiation facility	Excitation E_{exc} (MeV)	σ_{eff} (nb)	σ_{int} (mb eV)	Σ_{int} (μb keV)	$g\,\Gamma_0\,\Gamma_m/\Gamma$ (meV)	Lifetime τ (ps)	Refs
3 MeV LINAC	1.34±0.02			60±20			[65Ch]
2 MeV electron	0.74±0.01			2×10^{-2}	2.5×10^{-6}		[66Ca]
accelerator	1.12±0.01			0.1	3×10^{-5}		[69Bo]
	1.3±0.01			80±40/5	3×10^{-2}		
^{60}Co, 740 TBq	1.25		150±30				[60Yo]
^{60}Co, 48 TBq	1.1		82		8×10^{-3}		[63Ve1]
^{116}In^{m} reactor loop	<2.112	45					[66Ab]
^{60}Co, 7.8 PBq	1.3		80		3.5×10^{-2}		[73Za]
^{60}Co, 48 TBq	1.1		102±26				[74La]
^{60}Co, 18 TBq			350±40				[79Wa2]
^{60}Co, 18 TBq	1.33		110±20				[82Wa]
^{60}Co, 1.7 TBq	1.33		58±8				[82Kr]
^{64}Cu, 0.3–0.9 TBq	1.02	3.3±2					[83Vi]
^{60}Co, 74 TBq	1.33		140±10				[87Bi]
0.5–11 MeV, LINAC	4			9200±120			[91Ca2]
	6			25000±440			
4–15 MeV, LINAC	8			0.65±0.06 [a)]			[93Ma1]
	10			0.81±0.11 [a)]			
5–30 MeV, LINAC				0.51 [a)]			[93Sá2]
4 MeV LINAC			2.39±0.32 μb	16000±2400			[96La]
^{142}Pr, 1.5 TBq ;	1.4723					0.25	[97Ve]
	1.3269					11.5	
^{24}Na, 98 TBq	1.8425					2.8	

a) Calculated value in mb MeV

Cauchois and Boivin [66Ca], [69Bo] measured a very weak integral cross-section below 1.3 MeV at 0.74 and 1.12 MeV. Veres [63Ve1] carried out a calculation at the assumed 1.1 MeV energy of the activation level. Most authors adopted a value of 1.25 MeV energy without knowing the transition from the activation level to the isomeric state.

The dominance of non-resonant processes were reported for ^{111}Cd by Krcmar et al. [82Kr]. However, this observation was disproved by Bikit et al. [87Bi]. They used essentially the same technique for the measurement of the non-resonant contribution as Ref. [82Kr] and their result for the non-resonant cross-section was an order of magnitude smaller than that obtained in Ref. [82Kr] and it came to zero within one standard deviation error limit.

It is much more difficult, however, to observe the isomer ^{113}Cd^m with a 14-year half-life, and the papers cited above contain no information concerning this. So far there has been only one experiment with (γ, γ') reactions by ^{60}Co γ-rays [66Ve1]. To determine the cross- section of the isomer activity β counting was employed; this is because the metastable state of ^{113}Cd decays with a 99.86% branching ratio to the ground state of the nucleus ^{113}In via β-decay with an energy release of 0.58 MeV. The arrangement of the target and the source during the irradiation is shown in *Fig. 2.* The irradiation lasted two months, and the activity of the nuclear isomer ^{113}Cd^m with a half-life 14 years was investigated after the 49-minute ^{111}Cd^m had completely decayed.

The recalculated [97Ve] experimental (γ, γ') cross-section was 28±3 nb, since the 1.19 MeV activation energy level is higher than the 1.17 MeV γ-ray energy of the ^{60}Co source; this cannot contribute to photoactivation. The integral cross-section is 226±3 mb eV, and the half-life of the 1.19 MeV activation level is 7.1±0.7 ps.

The identification of the activation levels of ^{111,113}Cd was a long-standing problem. For the identification of all possible activation levels in ^{113}Cd, an (n, n'γ) experiment was carried out by Németh et al. [94Né2] and assigned an approximately 400 γ-transition. Based on their properties, three activation levels at 522.2 keV(7/2$^-$), 855.2 keV(5/2$^-$), and 1194.5 keV(3/2$^-$) energies are possible below 1.17 MeV. The 855.2 and 522.2 keV levels can only be populated by multistep photoactivation from the ground state. Under laboratory conditions the only available level is at 1194.5 keV because the likelihood of multistep excitation is very small.

From the nuclear structure point of view, the calculated (BXL)-reduced transition probabilities provide a good testing ground for theoretical nuclear structure calculations. The coupling between the even core and the odd nucleon in the intruder (opposite parity) orbital can be studied.

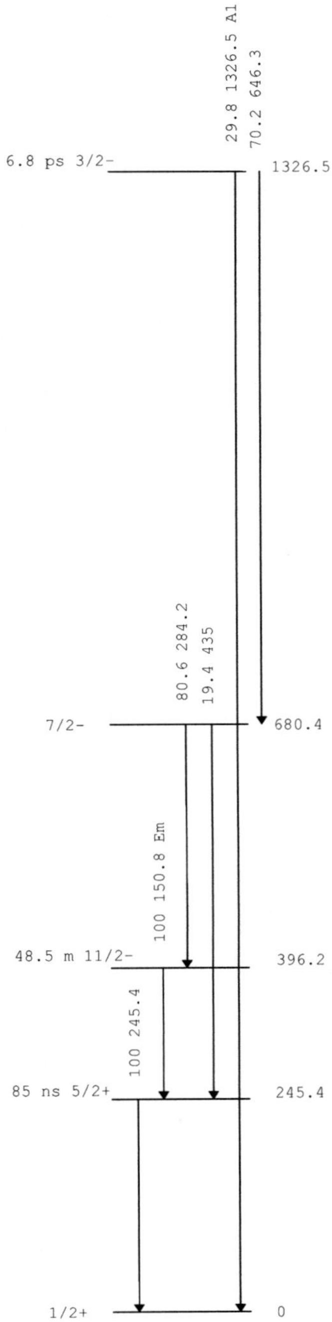

Fig. 49. Partial decay scheme of ^{111}Cd

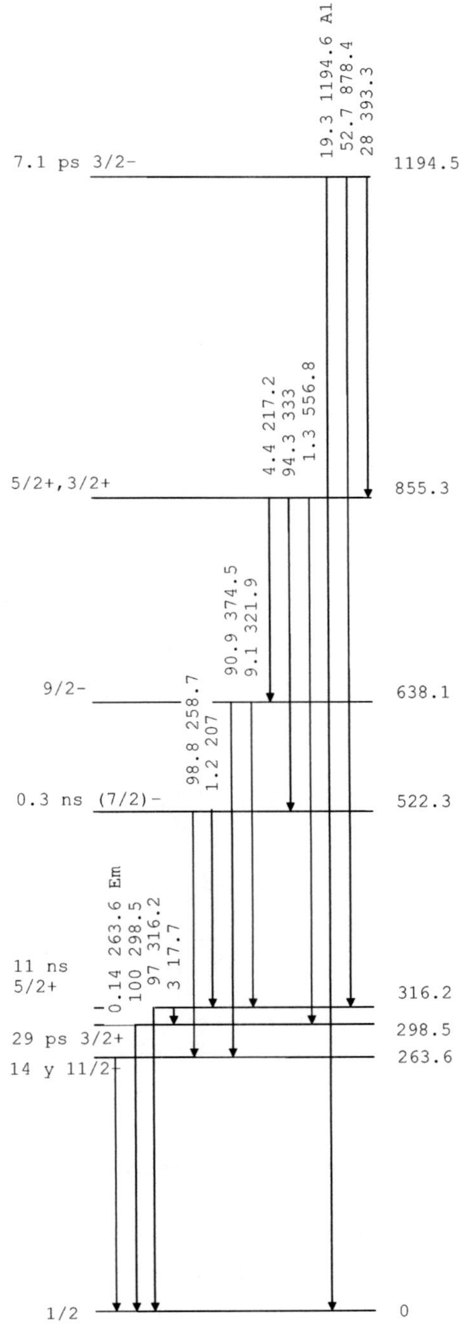

Fig. 50. Partial decay scheme of ^{113}Cd

However, in a stellar *s*-process scenario the low energy multistep excitations are more probable [91Kl] and thus the properties of lower energy levels can have more influence on the *s*-process behaviour of this nucleus.

On the basis of this $(n, n'\gamma)$ experiment Veres and Belgya (unpublished evalvation, 1977) suggested new activation levels. These are associated with the population of the isomeric states of ^{111,113}Cd. The proposed new activation levels were 1326.5 keV$(3/2^{-})$ for ^{111}Cd and 1194.5 keV$(3/2^{-})$ for ^{113}Cd. The branching ratios were determined and the lifetimes of the activation levels were updated using the correct decay schemes (see *Figs 49–50*). The cross-sections and other data in the higher energy range (1.5–50 MeV) are also listed in *Table 19* for ^{111}Cd.

Sáfár et al. [93Sá2] calculated the excitation function of the ^{111}Cd$(\gamma, \gamma')^{111}$Cd^{m} reaction, together with the available experimental data which are shown in *Fig. 51.* It shows a pronounced peak at the photoneutron threshold and then a sharp drop, so that the integrated cross-section is practically constant up to 30 MeV at least. The integral of the excitation curve for ^{111}Cd up to 15 MeV endpoint energy proved to be 0.51 mb MeV.

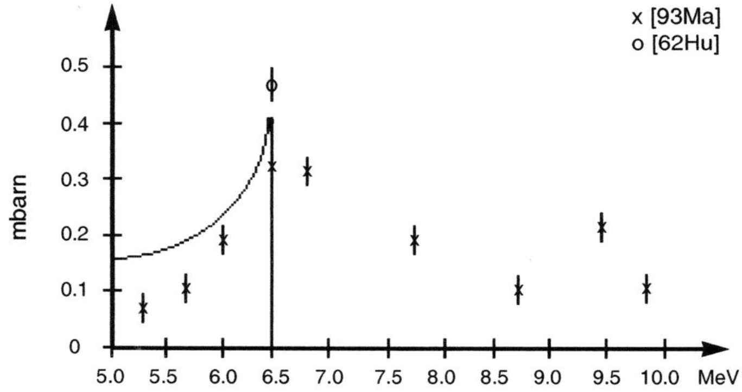

Fig. 51. Calculated excitation curve [93Sá] for ^{111}Cdm (solid line). Experimental data are given for comparison

6.11.2. Nucleosynthesis in the Cd In Sn region

One major group of nuclei remains quantitatively unexplained in the context of existing theories: there are a number of isotopes of heavy nuclei which cannot be formed under the typical conditions predicted for *s*-process or *r*-process neutron-capture synthesis. These bypassed or *pp*-process nuclei lie on the neutron-deficient side of the valley of beta stability, shielded by stable nuclei on their

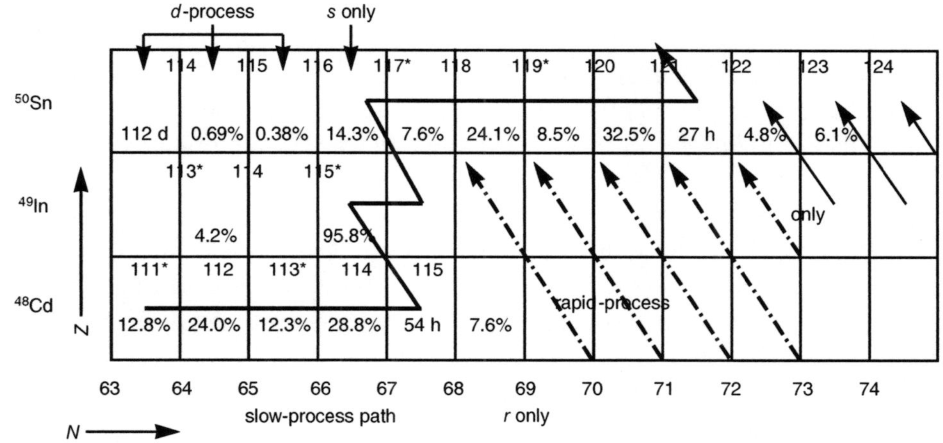

Fig. 52. The *s*-process path through the isotopes of cadmium, indium and tin

respective isobars from any contributions from neutron-capture mechanism. These abundance features are all consistent with the view that the bypassed nuclei are formed by proton-initiated reactions [*p*-process and perhaps a (γ, n) reaction]. Various alternative solutions have also been advanced, noted that both positron-capture and excited-state beta decays (photobeta reactions), proceeding at temperatures ~10^9 K, can transform *s* nuclei into *p* nuclei.

Audouze and Truran [75Au] studied the *p*-process in the 74–196 mass-region; in their calculations five individual cases gave rise to no enhancement of the nuclei which they believed to be the likely *p*-process candidates.

The difficulty with ^{113}In may be largely explained by the fact that it may be formed by the *s*-process as well, but a reliable estimate of the relative contributions is not available. A special consideration of the ^{113}Cd branch point is again dictated by the result of [75Au]. This indicates a severe underproduction of ^{113}In in their *p*-process calculations.

The calculation, neglecting the β-decay of ^{115}In gives:

$$N_s\left({}^{113}In\right)=\frac{(\sigma N)^{116}Sn}{\sigma\left({}^{113}In\right)\varsigma\left({}^{116}Sn\right)}\times$$
$$\times\left\{\varsigma\left({}^{115}Sn\right)\varsigma\left({}^{114}Sn\right)+\varsigma\left({}^{114}Cd\right)\varsigma\left({}^{115}In\right)\left[\frac{1}{f_-\left({}^{113}Cd\right)}-1\right]/\left[\varsigma\left({}^{113}In\right)\right]\right\}^{-1}. \tag{73}$$

Using the usual value $\tau_0 = 0.25$ and $\sigma(^{113}Cd) = 840$ mb, Ward et al. [76Wa] found that the total abundance of ^{113}In can be produced with $f_-(^{113}Cd) = 0.032$. Given the requirement that the solution of the average temperature be satisfied for consistency with the other branches, the correct branching ratio is produced at $T_8 = 3.3$.

Németh et al. [94Né2] studied the abundance contributions to the rare isotopes (^{113}In and ^{115}In) from the possible nucleosynthesis mechanism. They identified a possible activation level (1194.5 keV, 7.5 ps). It was confirmed that the main *s*-process is passing through ^{114}Cd, ^{115}In and ^{116}Sn with only minor branching at ^{113}Cd, ^{115}Cd and ^{115}In. The weak production of ^{113}In, ^{114}Sn and ^{115}Sn in the main nucleosynthesis process suggested that these abundances are dominant over other mechanisms *(Fig. 52)*.

6.12. ^{113}In and ^{115}In

There are two (^{113}In and ^{115}In) stable isotopes of indium with natural abundances of 4.3% and 95.7%, respectively. Their respective spins and parities are $9/2^+$ for the ground states and $1/2^-$ for the isomeric states. The respective energy levels and half-lives are 391.7 keV and 1.658 h for $^{113}In^m$, and 336.24 keV and 4.49 h for $^{115}In^m$.

The first conclusive evidence of nuclear isomerism in a stable nucleus was reported by Goldhaber, Hill and Szilard [38Go] in 1938. They found that the 4.1 h activity could be produced by fast neutrons but not by slow neutrons. This activity was assigned to a metastable state of ^{115}In excited by neutron impact and designated by In^{115*}. The X-ray excitation of the isomeric state of indium ($^{115}In^m$) was first observed by Pontecorvo and Lazard [39Po]. They produced $^{115}In^m$ by irradiation with X-rays of 1.85 MeV maximum energy from an impulse generator. Independently, Collins et al. [39Co], using X-rays from the Notre Dame electrostatic generator, found the threshold for the X-ray excitation to be 1.35±0.1 MeV. The latter authors [41Co] indicated that new information could be found about the character of the continuous X-ray spectrum produced by a monoenergetic electron beam, and some nuclear energy levels of the excited nucleus. Guth [41Gu], [45Gu] proposed an energy level diagram for indium in which the spin of the ground state is 9/2, the spin of metastable state is 1/2, and the spin of the activation state is 5/2. Waldman and Wiedenbeck [43Wa] and Wiedenbeck [45Wi] exemplified the excitation mechanism on indium and many other elements. Miller et al. [49Mi], [54Mi] verified the assumption that the excitation of $^{115}In^m$ is a line absorption. Dunworth and Pontecorvo [47Du], after irradiating indium with very intense X-rays, obtained the composite decay curve

Table 21. The excitation cross-sections and other data of ^{113,115}In nuclei

Irradiation facility	Excitation E_{exc} (MeV)	Isomers	σ_{int} (mb eV)	Σ_{int} (b MeV)	$g\,\Gamma_0\,\Gamma_m/\Gamma$ (meV)	$\sigma_{experiment}$ (nb)	Refs
Electrostatic gen.	1.04	^{115}In^m				0.1	[43Wa]
^{60}Co, 740 TBq	1.04	^{115}In^m	230 a)			29	[56Ik]
^{60}Co, 48 TBq	1.1	^{115}In^m	200		0.12	83	[63Ve1]
	1.04	^{113}In^m	106		2.8×10^{-2}	12	[66Ve3]
	1.04	^{115}In^m	434		0.12	49	
	1.078	^{115}In^m	105±27				[74La]
3 MeV Van	<0.9	^{113}In^m		$<5\times10^{-4}$	$<1.1\times10^{-4}$	<0.1	[67Bo]
de Graaff, energy	1.01			1.2×10^{-2}	3.2×10^{-3}	0.31	
resolution: ±10	1.13			7.5×10^{-2}	2.5×10^{-2}	1.4	
keV, beam	1.58			0.89	0.53	3.4	
intensity: 150 μA	0.829	^{115}In^m		$<3\times10^{-5}$	$<6\times10^{-6}$	<0.005	
	0.864			$<6\times10^{-5}$	$<1.2\times10^{-5}$	<0.005	
	0.935			8.5×10^{-3}	2×10^{-3}	0.18	
	1.08			11.5×10^{-2}	3.5×10^{-2}	0.18	
	1.45			0.59	0.32	2.8	
	1.57			0.25	0.16	2.1	
2 MeV electron	0.600	^{115}In^m		2×10^{-7}	1.8×10^{-8}		[69Bo]
accelerator	0.830			1.3×10^{-7}	2.3×10^{-8}		
	0.935			0.02	4.5×10^{-3}		
	1.070			0.3	9×10^{-2}		
	1490			2	1.2		
^{116}In^m reactor loop	<2.112	^{115}In^m				123	[66Ab]
^{46}Sc, 16TBq	1.078	^{115}In^m				7	
^{142}Pr, 15TBq	<1.57					97	
^{24}Na, 98 TBq	<2.75					135	[73Ve2]
^{22}Na, 260 GBq β^+	1.078	^{115}In^m				0.1–1 barn	[72Mu]
^{60}Co, 7.8 PBq	1.13	^{113}In^m	52.6				[73Za]
	1.078	^{115}In^m	90				
^{64}Cu, 14 GB q/cm^2	1.13	^{113}In^m				7.3 b)	[79Vi]
β^+	1.078	^{115}In^m				11 c)	
^{60}Co, 18.5 TBq	1.078	^{115}In^m	230				[82Wa]
^{60}Co, 25 TBq	1.078	^{115}In^m	53.9 d)			380	[81Lj]
^{60}Co 3 PBq	1.078	^{115}In^m	181				[86Yo]
^{60}Co, 0.1 MBq	1.13	^{113}In^m				81 μb e)	[87Kr]
2–5 MeV S-D c)	3.2	^{115}In^m		7.6±1.2			[92Neu]
Four accelerators'	4	^{113}In^m		13 mb keV			[91Ca2]
0.5–11 MeV	6			47 mb keV			
	4	^{115}In^m		18 mb keV			
	6			67 mb keV			
4–15 MeV, BS	8			2.1 mb MeV			[93Ma1]
	10			4.2 mb MeV			
15–50 MeV, BS	*			9.3 mb MeV			[91Sc]
4 MeV, 25 A	4 MeV	^{113}In^m		43			
LINAC	endpoint	^{115}In^m		48			[95La2]

a) [60Yo]; b) $\sigma_{res,exp}$ = 4.8 barn and $\sigma_{res,teor}$ = 0.6 μb; c) $\sigma_{res,exp}$ = 1.9 b; d) σ_{nonres} = 53.9 mb eV, according to the measurement of [86Yo] it is less than 3.3% of the 181 mb eV value, while it is no more than 6 mb eV; e) σ_{res} = 81 and σ_{nonres} = 20 nb

of the two isotopes, from which the 105 minute activity was subtracted leaving the 4.4-hour activity. Chertok and Booth [65Ch] determined the photoexcitation of ^{115}In as a function of electron energy, and Booth and Brownson [67Bo], [71Bo] as a function of photon energy.

Up to 1954 attempts to produce metastable states using monoenergetic γ-rays of nuclear origin for activation all failed. Goldhaber et al. [39Go], for instance, carried out an experiment with γ-rays emitted by 0.5 g ^{226}Ra to obtain the metastable level of indium by the $^{115}In(\gamma, \gamma')^{115}In^m$ reaction. The negative results of their experiment are attributed by Guth [41Gu] to the fact that the γ-energies from ^{226}Ra do not correspond to the possible excitation levels of indium. Goldemberg and Katz [53Go] measured the cross-section of $^{115}In^m$ induced photonuclear reaction in indium by the X-rays from 25 MeV betatron. In 1954 Harbottle produced isomeric activity with γ-rays from ^{60}Co sources with activities of 6.3 TBq (170 Ci) to 66.7 TBq (1800 Ci); and with γ-rays of ^{182}Ta sources with activity of 48 TBq (1300 Ci) [54Ha]. From the produced isomer activity data he made estimations for the overall cross-section nano-barn (nb) order of magnitude. Ikeda and Yoshihara [56Ik], and then Veres [63Ve1], reported 29 nb and 48 nb values, respectively. Yoshihara and Veres, independently of each other, studied the Compton-scattered intensity [60Yo], and some calculations were carried out [63Ve1]. From these it was determined that the number of scattered γ'-quanta in one eV energy interval from a ^{60}Co source of 18.5 TBq (500 Ci) activity is the same order of magnitude as the number of *bremsstrahlung* quanta of an accelerated electron beam of 100 μA intensity.

The observed results were used to estimate the respective spins, parities, and the level width of the excited states *(Figs 53, 54)*. The photoactivation of nuclear isomers can also be used in a variety of fields: e.g., dose rate measurements of high activity radiation sources, hot-atom chemistry, non-destructive isomer activation analysis, and as reference cross-section values in the data evaluation of many photoexcitation experiments.

With respect to the great number of publications in *Table 21,* those papers giving similar results are omitted. In the early photoexcitation experiments, due to the lack of exact gamma-decay schemes, the integral cross-section values are based on estimated values of the activation level energy. This suggests that it would be useful to re-evaluate early data.

Cascade γ-ray calculations were performed to determine the cross-section in the 2 to 23 MeV energy range. Branching ratios of each step were determined using the conventional optical statistical model. E1 transitions only were considered and the results of the calculation are shown in *Fig. 55* [93Sá2]. Similar calculations were performed and similar results were obtained for the $^{103}Rh(\gamma, \gamma')^{103}Rh^m$ and $^{111}Cd(\gamma, \gamma')^{111}Cd^m$ reactions.

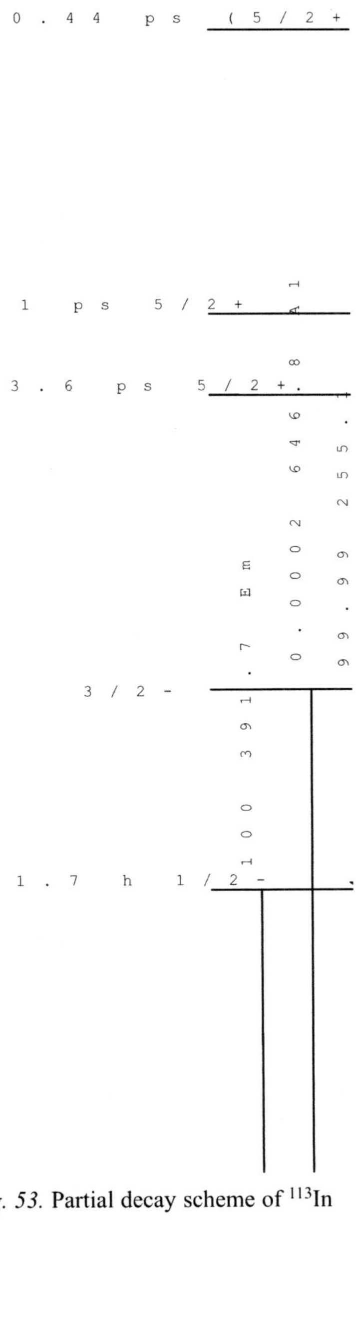

Fig. 53. Partial decay scheme of ^{113}In

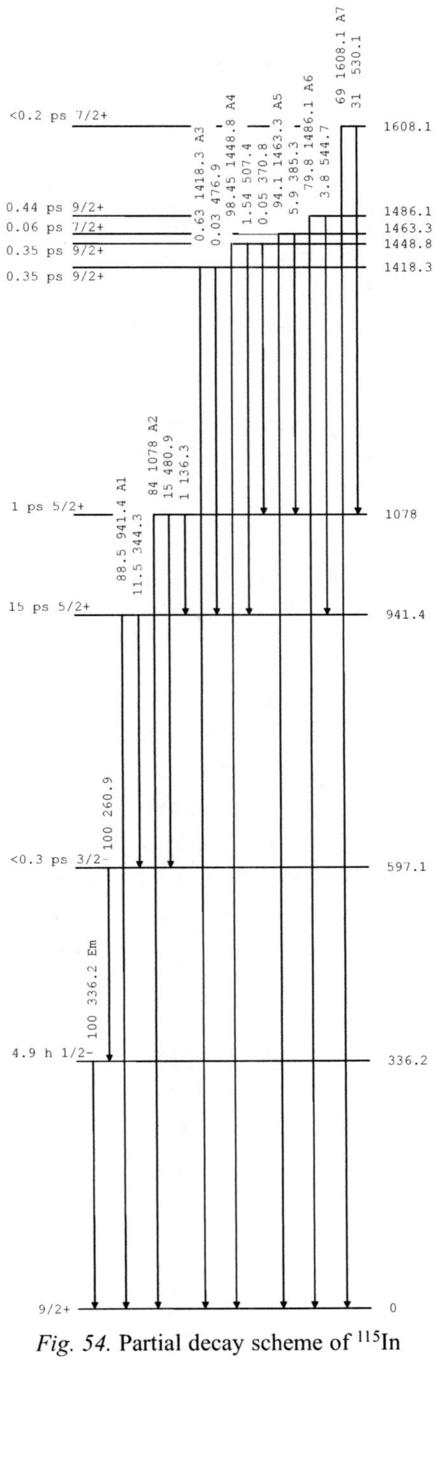

Fig. 54. Partial decay scheme of ^{115}In

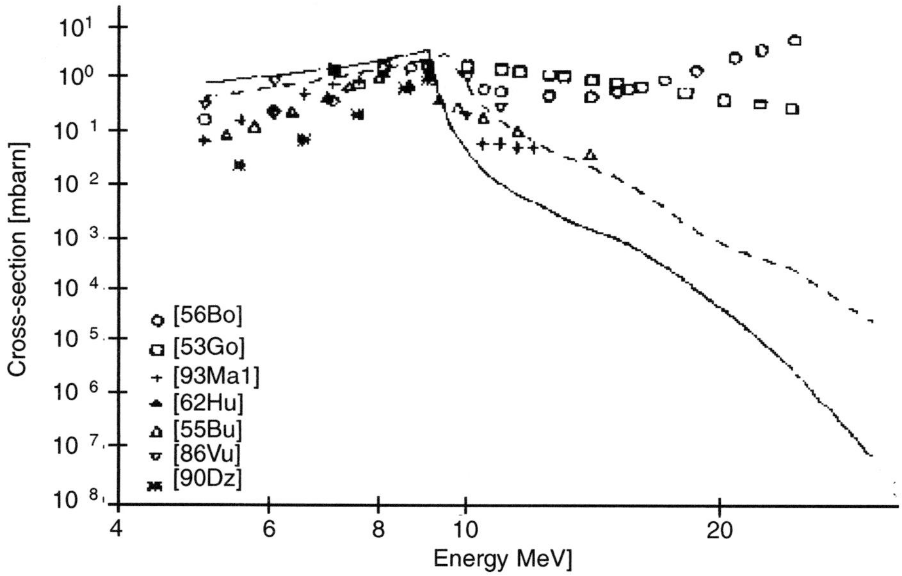

Fig. 55. Calculated excitation functions for $^{113}In(\gamma, \gamma')^{113}In^m$ (dashed line) and for $^{115}In(\gamma, \gamma')^{115}In^m$ (solid line). Experimental data from earlier measurements (Refs marked) are given for comparison

The calculated saturated values of the integral cross-section proved to be 8.9 and 9.1 mb MeV for ^{113}In and ^{115}In, respectively. The difference in the excitation curves of the indium isotopes above a (γ, n) threshold is a consequence of the fact that the applied systematics of the level-density parameter reflect the $2d_{1/2}$ sub-shell closure. The large variation in their size, found in accordance with prior knowledge, calls for further explanation.

6.13. ^{117}Sn and ^{119}Sn

There are ten natural isotopes of tin, two of them having isomers (^{117}Sn and ^{119}Sn) with spins $1/2^+$ in their ground states and $11/2^-$ in their isomeric states. The energies and half-lives of the isomeric states are 314.58 keV; 13.6 d and 89.531 keV; 293.1 d, respectively. The natural abundance is 7.68% for ^{117}Sn and 8.59% for ^{119}Sn.

There have been very few photoexcitation experiments with the $^{117,119}Sn$ stable isotopes. In 1963 Kaminishi and Kojima [63Ka] irradiated 22 stable elements with *bremsstrahlung* from a 6 MeV linear electron accelerator having a beam current of 20–100 mA (peak), 20–100 μA (average), and a pulse width 5

μseconds. It was confirmed that 18 isomers were directly produced from the stable nuclides with photons. Among these isomers, the isomeric activity of $^{117}Sn^{m}$ was also observed. Detection of the low isomer yields from ^{113}In and ^{135}Ba was not reliable and no isomers of even-even nuclides were observed in the experiment. Designing and constructing an automatic apparatus, the authors were able to detect a fraction of μg or less in the case of 15 elements. However, the sensitivity in the case of ^{117}Sn is lower by three or four orders of magnitude.

Ramchandran and McIntyre [69Ra] used neutron-capture γ-rays to excite one nuclear energy level in Sn in the 5–9 MeV energy region. Angular distribution measurements of the scattered γ-rays showed that the transitions were all consistent with dipole radiation. The ^{117}Sn and ^{119}Sn have a neutron-binding energy which is less than that of the 6.988 MeV excitation; thus the scattering measured represents only the elastic scattering of the 6.988 MeV γ-ray.

Abrams and Pelekis [66Ab] irradiated a tin target with γ-rays of ^{116}In radioactive source, produced in a reactor loop; they got no isomeric activity from the ^{117}Sn and ^{119}Sn nuclei. Veres et al. [73Ve2] irradiated 17 elements having isomeric states, among them a tin target too, with γ-rays of four (^{46}Sc, ^{60}Co, ^{142}Pr and ^{24}Na) radioisotopes. No measured isomeric activity was observed after the irradiation by ^{46}Sc, ^{60}Co, ^{142}Pr radiation sources, but with γ-rays (E_{max} = 2.75 MeV) of ^{24}Na, $^{117}Sn^{m}$ and $^{119}Sn^{m}$ were produced with overall cross-sections of 56 nb and 175 nb, respectively.

Carroll et al. [91Ca2] examined the photoexcitation of 19 nuclides, among them Sn as well. The integral cross-section value of ^{117}Sn was measured as 3.20±0.47 mb keV at 4 MeV, and 8.80±0.26 mb keV at 6 MeV endpoint irradiation energy. Lakosi et al. [93La1], also studied the excitation of isomers around and above the neutron-binding energy, and estimated saturation values as 3.2 mb MeV for ^{117}Sn and 2.3 mb MeV for ^{119}Sn above 7–9 MeV. In Ref. [95La2] the photoactivation of 25 isomers by 4 MeV endpoint energy *bremsstahlung* was studied. The effective cross-sections were evaluated. The integral cross-sections were also determined, based on the Brink–Axel hypothesis. This used the Lorentzian extrapolation of the photoabsorption cross-section for a reference given for giant dipole resonance by Dietrich et al. [88Di]. The results demonstrate that the isomer activation is a multiparametric process. The integrated cross-section value for 4 MeV endpoint energy is 4.4±0.66 μb MeV and the effective cross-section value for a flux >1 MeV is 0.655±0.087 μb for the ^{117}Sn nucleus [96La].

6.14. ^{123}Te and ^{125}Te

There are eight natural isotopes of the element tellurium, two of them having isomers (^{123}Te and ^{125}Te) with spins $1/2^+$ in their ground states and $11/2^-$ in their isomer states. The energies and half-lives of the isomeric states are 247.5 keV; 119.7 d and 144.73 keV; 58 d, respectively. The 247.75 keV isomeric state of ^{123}Te^{m} moves to the ground state via levels at 159 keV and 88.5 keV. The natural abundance is 0.89% for ^{123}Te and 7.14% for ^{125}Te.

The photoexcitation experiments for the stable isotopes ^{123}Te and ^{125}Te are also very scarce. Zaparov et al. [73Za] carried out an experiment to produce isomeric activity in a natural tellurium target by irradiating it with ^{60}Co γ-rays having 6×10^{12} γcm^{-2} s^{-1} intensity at the place of irradiation. They measured the 160±5 keV γ-line of ^{123}Te^{m} with the half-life of 103±5 d, and the integral cross-section was estimated as being about 6 mb eV in the 1.17–1.33 MeV energy range and about 2.7 mb eV in the 1.0–1.17 MeV energy range. However, for ^{125}Te no isomer activity was measured because the 110 keV γ-transition has a very high conversion coefficient and thus it was not possible to determine precisely the intensity of the characteristic X-rays of 31 keV; this is why a threshold value of ≤ 0.16 mb eV was given.

Veres et al. [73Ve2] excited the ^{123}Te and ^{125}Te isomers by inelastic resonance scattering of Compton-scattered γ-rays from a high activity ^{24}Na radioactive source. The overall cross-sections are 520 mb for ^{123}Te^{m} and 870 nb for ^{125}Te^{m}, provided that the activation levels lie in the 1.33–2.75 MeV energy range.

Carroll et al. [91Ca1] determined the integral cross-sections of ^{123}Te^{m} and ^{125}Te^{m} by different irradiation energies of *bremsstrahlung*, which are: for ^{123}Te^{m} at 1±0.5 MeV, $\Sigma_{int} < 0.6\pm0.2$ μb MeV; at 2.8±0.2 MeV, $\Sigma_{int} = 20\pm3$ μb MeV; at 4.2±0.2, $\Sigma_{int} = 70\pm10$ μb MeV, and for ^{125}Te^{m} at 4.2–4.5 MeV, $\Sigma_{int} = 70\pm30$ μb MeV. The authors [91Ca2] measured the integral cross-sections of 19 isomers at irradiation energies of 4 MeV and 6 MeV, as already mentioned above. Their experimental results for ^{123}Te were 42±7.3 μb MeV at 4 MeV irradiation energy and 68±3.2 μb MeV at 6 MeV irradiation energy.

Lakosi and Sáfár [93La1] studied the photoexcitation of isomers around and above neutron binding energy (7–9 MeV). The authors estimated the saturation integral cross-sections of 33 isomers, among them, also ^{123}Te^{m} and ^{125}Te^{m} as 4.3 mb MeV and 3.6 mb MeV, respectively.

6.15. ^{129}Xe and ^{131}Xe

There are eight natural isotopes of the element xenon, two of them having isomers (^{129}Xe and ^{131}Xe) with spins $1/2^+$ and $3/2^+$ in their ground states, respectively; and both with spins of $11/2^-$ in their respective metastable states. The energy levels and half-lives of these isomers are 236.1 keV; 8.88 d for ^{129}Xe, and 163.9 keV; 11.92 d for ^{131}Xe, and their rates for natural abundances are 26.4% and 21.2%, respectively.

As can be seen from this compilation, a number of experiments have been published on isomer production of stable isotopes by (γ, γ') reaction. But till now only two papers [90Ta] and [95La2] are known in the literature about the photoactivation of ^{129}Xe^m and ^{131}Xe^m. It can be presumed that one of the reasons is the difficulty in handling the xenon target in its gas form.

In the first experiment Tam et al. [90Ta] used a natural xenon gas target of 17.6 grams; this was irradiated with a 7.4 PBq (200 kCi) ^{60}Co radiation source. The initial isomeric activity of the xenon target was 0.56 counts per sec and 0.054 count per sec at 196 keV and 164 keV, respectively. The measured half-life data were 8.85±0.04 d for ^{129}Xe^m and 11.9±0.2 d for ^{131}Xe^m. The values for the integral cross-sections for the isomer production by photoexcitation of ^{129}Xe and ^{131}Xe at the supposed activation levels around 1 MeV were evaluated as 75±20 mb eV and 25±8 mb eV, respectively.

It is difficult to follow the activation mechanism from the level scheme of ^{129}Xe because no direct or cascade transitions were identified from the excited levels to the isomeric state (see, for example [78Pa] and [79Ir]). Zhao et al. [88Zh] found transition to the 236.1 keV($11/2^-$) isomeric state only from the level of a high spin-value (15/2, 13/2) range. It is very unlikely that these states can be excited from the $1/2^+$ ground state.

Among the established levels of ^{131}Xe [79Ir], [83Lö], [96Fi] there are four or five levels from which both the $3/2^+$ ground state and the $11/2^-$ isomeric state are fed. On the basis of these data a possible partial photoexcitation and γ-decay scheme can be made *(Fig. 56)*. Nevertheless, in the case of the activation level A1 (and knowing level lifetime and branching ratios), the calculated integral cross-section was two orders of magnitude less than that deduced from the measurement. For the second activation level (722.9 keV, $5/2^+$) the parameters were again known and the authors [90Ta] calculated them six order of magnitude less than the measured value. From the 973 keV, $7/2^+$ level a transition is seen to the ground state; this feeds the isomeric state too, but the branching ratio was found to be very low. Since it is evident that activation levels exist even below 1.33 MeV, further investigations are needed to find new transitions between established levels or to introduce new levels into the level schemes.

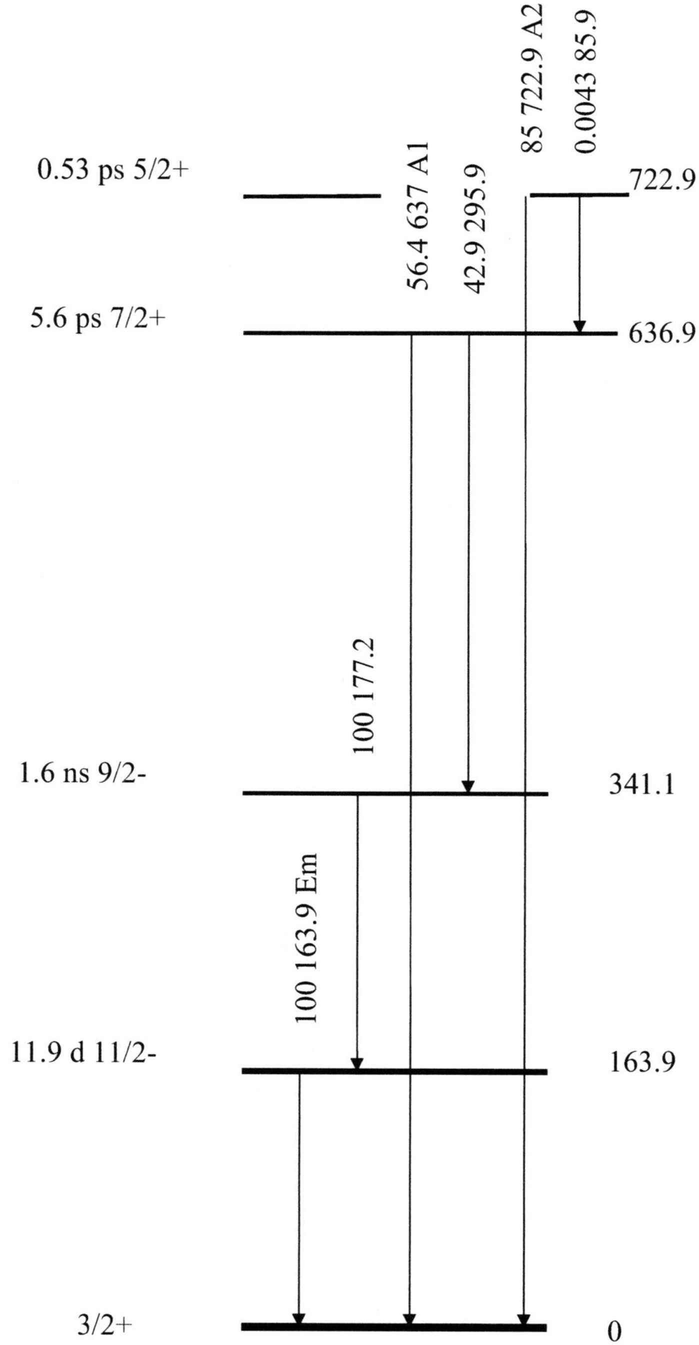

Fig. 56. Partial decay scheme of ^{131}Xe

In the higher energy range a series of isomers were excited by a (γ, γ') reaction, using *bremsstrahlung* of a 4 MeV, 25 μA linear accelerator [95La2]. The effective overall cross-sections for a flux >1 MeV at 4 MeV endpoint *bremsstrahlung* energy are 14 μb for ^{129}Xe and 22 μb for ^{131}Xe.

6.16. ^{135,136,137}Ba

There are seven natural isotopes of the element barium, five with a spin of zero and two with a spin of $3/2^+$ in the ground states. The three isotopes indicated above have isomer states of $11/2^-$, 7^- and $11/2^-$, respectively. The energy and half-lives of the isomeric states are: 268.3 keV; 28.7 h for ^{135}Ba, 2030.6 keV; 0.3 sec for ^{136}Ba and 661.66 keV; 2.55 minutes for ^{137}Ba. The natural abundances of these isotopes are 6.592%, 7.854% and 11.23%, respectively.

In the literature, there is no evidence for the isomer of the even-even nucleus $_{56}$Ba$_{80}$ being observed by photoexcitation. The results were similar to the *bremsstrahlung* experiment, [63Ka] with no measured isomer activity for five even-even nuclides [73Ve3]. This reason does not seem to be the case, considering the fact that a trace of activity could be detected. At present there is no information to account for this result. That is, in order to produce the isomer from a nuclide, the nuclide must first be excited to a higher-lying level than the isomeric one, and then the transition to the isomeric level (not to the ground) must be allowed with reasonably large probability. This is decided by the respective natures of the two levels. Little is known about the effective transitions leading to the isomeric level due to a lack of information about the nature of highly excited levels. This is especially true for even-even nuclides with large mass number, and these are substantially related to this problem. The lack of success in producing even-even nuclides through photoexcitation process can largely be ascribed to the structures of these nuclides and to the interactions with photons.

For ^{135,137}Ba^m the integral cross-section values were measured by Zaparov et al. [73Za] with γ-rays from a ^{60}Co radiation source. The result for ^{135}Ba^m is 3.13 μb eV, if the activation level lies in the 1.17–1.33 MeV energy range and 1.4 μb eV, if the activation level lies in the 1.0–1.17 MeV energy range. The authors determined the integrated cross-sections for ^{137}Ba, too, and its value was $\leq$4 μb eV related to the energy range of 1.17–1.33 MeV.

In the higher energy range Carroll et al. [91Ca2], [93Ca1], and Lakosi and Sáfár [93La1], [95La2], investigated the photoexcitation of $^{135, 137}$Ba isomers.

Mazur et al. [93Ma1] studied the cross-sections of γ inelastic scattering with excitation of the isomeric state of the ^{137}Ba nucleus. It was measured at

Table 22. The excitation cross-sections and other data for $^{135,137}Ba$ nuclei

Irradiation facility and/or intensity	Excitation E_{exc} (MeV)	Isomers	σ_{int} (mb eV)	Σ_{int}	$g\,\Gamma_0\,\Gamma_m/\Gamma$ (meV)	$\sigma_{exper.}$ (μb)	Refs, remarks
3.5×10^6 γ/cm^2 s eV	1.17–1.33	$^{135}Ba^m$	3.13		1.27×10^{-3}		[73Za]
	1.0–1.17			1.4	4.4×10^{-2}		(^{60}Co)
7.8×10^6 γ/cm^2 s eV	1.17–1.33	$^{137}Ba^m$	$\leq 4\times10^{-3}$				
3.5×10^6 γ/cm^2 s eV							
0.5–11 MeV,	4	$^{135}Ba^m$		13±0.62 mb keV			[91Ca2]
LINAC	6			60±1.1 mb keV			
	4	$^{137}Ba^m$		3.3±0.023 mb keV			
	6			19±0.23 mb keV			
	2.0			<0.15 mb keV			[93Ca1]
	3.2±0.1			2.2±0.2 mb keV			
	4.5±0.2			85±8 mb keV			
4–15 MeV,	8	^{137}Ba		1.13±0.1 mb MeV			[93Ma1]
M30 Microtron	10			1.2±0.15 mb MeV			
Estimated	> 7–9	^{135}Ba		6.9 mb MeV			[93La1]
saturation		^{137}Ba		6.7 mb MeV			
4 MeV, LINAC	4 endpoint	^{135}Ba		36 μb MeV		5.3	[95La2]
		^{137}Ba		6.2 μb MeV		0.9	

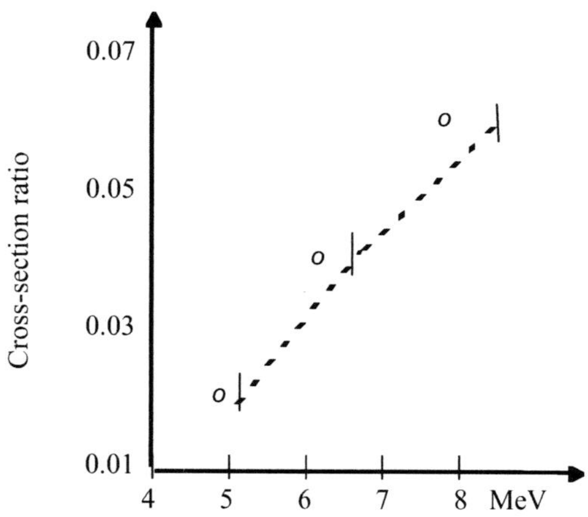

Fig. 57. The experimental isomer cross-section ratio as the function of the energy

4–15 MeV energies in ΔE = 0.5 MeV steps. For $(\gamma, \gamma')^m$ reactions below the (γ, n) threshold the energy dependence of the isomer cross-section ratio σ_m/σ_{tot} is considered *(Fig. 57)*.

Carroll et al. [93Ca1] analysed the measured data of ^{137}Ba because the integrated cross-section quantity is of primary importance for photoactivation studies. The normalised activation, Y_f, which gives the fractional yield of the isomer produced by an irradiation per photon incident, is:

$$Y_f = \frac{N_{\text{iso}}}{N_T \Phi_0} = \sum_j (\sigma\Gamma)^j_{\text{iso}} \; F(E_j, E_0), \tag{73}$$

where N_{iso} is the number of isomers produced, N_T is the number of target nuclei, Φ_0 is the total photon flux for the specified endpoint energy E_0, and $F(E_j, E_0)$ is the distribution of energies within the *bremsstrahlung* continuum, normalised to 1.

The summation in equation (73) allows the participation of a plurality of well-spaced gateways. *Fig. 58* displays the excitation function of the production of ^{137}Ba^m. For comparison, the asterisks represent the experimental results obtained from *Table 22*. The good agreement observed between values measured in completely different experimental environments established confidence in the procedures. Intermediate states were identified by sharp increases in the normalised activation.

Lakosi et al. [93La1], [95La2] studied the excitation of isomers around and above neutron binding energy and estimated the integral cross-section for ^{135}Ba and ^{137}Ba as 6.9 mb MeV and 6.7 mb MeV, respectively. In the experiment of [95La2] the authors excited a series of isomers with a (γ, γ')-reaction using *bremsstrahlung* from a 0.9 mm platinum converter placed against the electron beam of a 4 MeV, 25 μA linear accelerator. They measured the effective

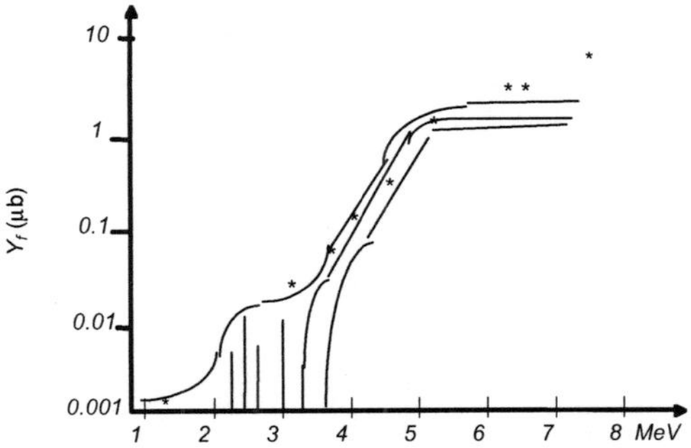

Fig. 58. Normalised yield, Y_f as function of electron energy for the population of ^{137}Ba

cross-section for a flux >1 MeV, and the integrated cross-sections for 4 MeV endpoint energy. These results are 5.3 μb and 36 μb MeV for ^{135}Ba, and 0.9 μb and 6.2 μb MeV for ^{137}Ba, respectively.

It is difficult to make comparisons between the experimental results of *Table 22.* because they are concerned with different energy ranges.

6.17. ^{167}Er

There are six stable isotopes of the element erbium, five with spin zero and only one (^{167}Er) with a spin $7/2^+$ in the ground state; the natural abundance of this isotope is 22.95%. The ^{167}Er isotope has an isomeric state with a spin of $11/2^+$, energy of 207.8 keV and with a half-life of 2.269 seconds.

In 1957 Hammer and Stewart [57Ha] reported that the radioactivity with a half-life of 2.5 seconds in natural erbium with *bremsstrahlung* using 75 MeV synchrotron would be due to one of three isomers: ^{165}Er^m, ^{167}Er^m, and ^{169}Er^m. Mihelich and Harmatz [57Mi] determined that the radioactivity was due to ^{167}Er^m. Kaminishi and Kojima [63Ka] irradiated 17 elements with *bremsstrahlung* of 4–6 MeV of a linear accelerator and they measured (with a G–M counter) the highest values of isomeric activity (4.9×10^7 counts per minute) of

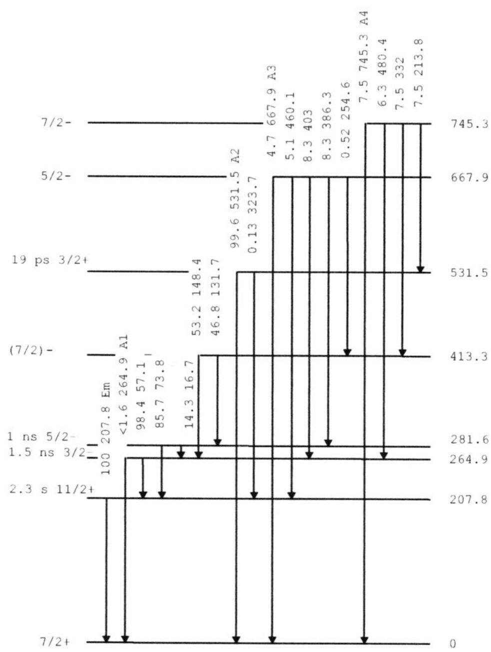

Fig. 59. A partial excitation and decay scheme of the ^{167}Er nucleus

Table 23. The excitation cross-sections and other data of the ^{167}Er nucleus

Irradiation facility and/or intensity	Excitation energy (MeV)	σ_{int} (mb eV)	Σ_{int}	$\sigma_{experiment}$ (μb)	Refs
^{116}In^{m} reactor loop	<2.112			0.75±0.19	[66Ab]
^{24}Na, 3.7 TBq	<2.75			250±50	[69Ab]
3.5×10^6 γ/cm^2 s eV	1.17–1.33 (^{60}Co)	≤20.5			[73Za]
^{142}Pr, 18 TBq	<1.57			1.2±0.4	[73Ve2]
^{24}Na, 90 TBq	<2.75			214±43	[73Ve2]
0.5–11 MeV LINAC	4		46±1.1 mb keV		[91Ca2]
	6		340±3.6 mb keV		
4–15 MeV, M30	8		10.5±1.2 mb MeV		[93Ma1]
Microtron	10		12.0±2.2 mb MeV		
Estimated saturation	>7–9		13.5 mb MeV		[93La1]
4 MeV LINAC	4 (endpoint)		3.56 mb MeV		[95La2]

^{167}Er. Abrams and Pelekis [66Ab] irradiated an erbium target with γ-rays (E_{max} = 2.112 MeV) of ^{116}In^{m} produced in a reactor loop. The overall cross-section was 0.75±0.19 μb. Abrams and Lakosi [69Ab], with the X-rays (E_{max} = 2.75 MeV) of ^{24}Na, determined a value of 250 ±50 μb. Zaparov et al. [73Za] irradiated with ^{60}Co sources and reported for the integral cross-section a value of ≤20.5 mb eV in the energy range of 1.17–1.33 MeV. The cross-sections and other values can be found in *Table 23.*

From the excitation and decay scheme of *Fig. 59* it can be seen that at least three activation levels exist below 1 MeV. From these direct and cascade transitions can populate the isomeric state. However, their respective widths are probably too narrow and/or their branching ratios are too small to bring about measurable isomeric activity.

6.18. ^{176}Yb

There are seven natural isotopes of the element ytterbium and only the ^{176}Yb nucleus has an isomeric state. It is an even-even nucleus with a spin of 0^+ in the ground state and (8^-) in the isomeric state. The energy of the isomeric state is 1050.0 keV, and the half-life is 11.4 seconds. The natural abundance of this nucleus is 12.7%.

Up until now no experiment dealing with the photoexcitation of ^{176}Yb^{m} has been reported. Not much is known about the effective transitions leading to the isomeric level due to a lack of information on the respective natures of highly excited levels. Therefore it can be noted that failure in producing even-even nuclides via the photoexcitation process can in a large part be substantially ascribed to the respective structures of these nuclides and to the interactions with photons.

6.19. ^{176}Lu

There are two stable isotopes of the element lutetium and their rates of natural abundance are 97.41% (^{175}Lu) and 2.59% (^{176}Lu), respectively. The latter quasi-stable nucleus has an isomeric state with energy and spin of 123 keV(1^-) and the half-life of the isomeric state is 3.635 hours.

The isomer of the ^{176}Lu nucleus was examined in 1946 by Dunworth and Pontecorvo [47Du] with the aid of 2 MeV X-rays of an electrostatic generator. On the basis of the reactions ^{115}In(γ, γ')^{115}In^{m} and ^{176}Lu(γ, γ')^{176}Lu^{m}, it was found that the activation cross-section was of one order of magnitude less for ^{115}In than for ^{176}Lu, i.e., $\sigma_{In}/\sigma_{Lu} = 0.1$.

6.19.1. Experimental and calculated data

The level scheme of ^{176}Lu, a naturally occurring radioactive nuclide (half-life 3.6×10^{10} years), exhibits a rotational band structure [81De], [88Kä]. As inter-band *K*-forbidden transitions are highly retarded, the $I^{\pi} = 1^-$, $K = 0$ isomeric state at 123 keV decays completely to ^{176}Hf by b^- emission, but does not decay to the 7^- ground state of ^{176}Lu^{g} (as indicated in *Fig. 60*). Watanabe et al. [81Wa]

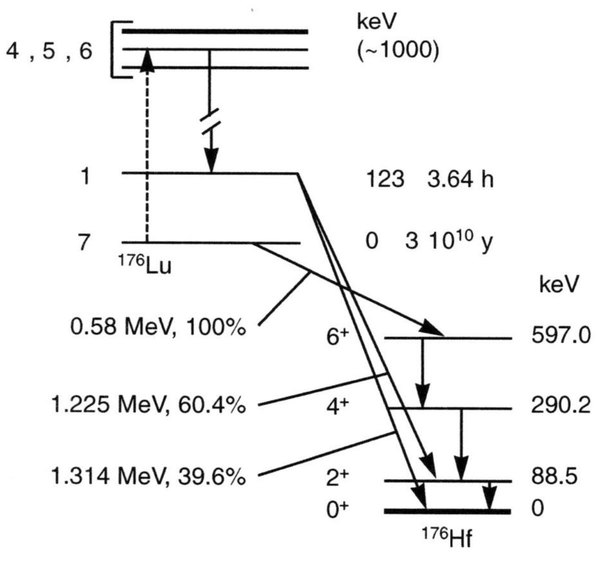

Fig. 60. A partial decay scheme of ^{176}Lu.
Data are taken from [^{78}Le]

supposed that the three higher levels in the figure (~1000 keV; 4^-, 5^-, 6^-) are 1083, 1046 and 946 keV, each with spins of 5. The existence of activation levels having non-zero transition probabilities for both the ground and isomeric states have been verified by many photoactivation experiments using ^{60}Co sources. However, no measured isomer activity was found after the irradiation of the lutetium target by γ-rays (662 keV) of the ^{137}Cs source [84No], [92La] *(Table 24).*

Table 24. The excitation cross-sections of the ^{176}Lu nucleus

Irradiation facility and/or intensity	Excitation energy (MeV)	σ_{int} (mb eV)	Σ_{int} (mb keV)	$\sigma_{experiment}$ (nb)	Refs, remarks
^{60}Co, 740 TBq	1–1.17; 1.17–1.33			1000–2400	[60Yo]
^{60}Co, 6; 740; 3000 TBq	1–1.17; 1.17–1.33			23–47	[70Ve]
(3.5–7.8)×10^6 γ/cm^2 s eV	1–1.17; 1.17–1.33	2.1– 4.7			[73Za]
^{64}Cu , 0.63 TBq	1.083			26000±9000	[81Wa]
^{137}Cs, 55.5 TBq	0.662;			<10	[85No]
^{60}Co, 122 TBq	1–1.33			38±10	
^{137}Cs, 56 TBq	0.662				[95La1], no I_{iso}
^{60}Co, 3000 TBq	~1	62±20		40±13	
0.5–11 MeV LINAC	6		140±1; and 350±2.3		[91Ca2]
Estimated sat. value	>7–9		5200		[93La1]
4 MeV, 25 μA LINAC	4 (endpoint)		310±47	45900±5900	[96La]

Several attempts have been made to produce the isomeric state of ^{176}Lu using large ^{60}Co γ-ray sources. The results have varied: Yoshihara [60Yo] – 1.0–2.4 μb; Veres and Pavlicsek [70Ve] – 23–47 nb. It depends on whether the activation level lies in the energy ranges of 1.00–1.17 MeV or 1.17–1.33 MeV. To resolve the discrepancy and to investigate the photoactivation cross-section of ^{176}Lu with lower energy photons, Norman et al. [84No] irradiated a natural lutetium target with large photon fluxes from both ^{137}Cs (E_γ = 662 keV) and ^{60}Co sources. Following the ^{137}Cs bombardment, no excess count above the ^{167}Lu background was observed, and an upper limit on the photoactivation cross-section of 10 nb for 662 keV photons was established. After ^{60}Co irradiation, assuming that both the 1173 keV and 1332 keV photons contribute to the observed yield, a photoactivation cross-section of 38±10 nb for ^{60}Co photons was inferred. This result is in good agreement with that of Veres and Pavlicsek, while it disagrees strongly with that of Yoshihara. As was discussed by Veres and Pavlicsek, the origin of this discrepancy does not have an immediate explanation.

Watanabe et al. [81Wa] studied the nuclear excitation process of ^{176}Lu by positron annihilation. The authors used a positron source of 0.63 TBq of ^{64}Cu for irradiation of a natural lutetium foil (20 mm diameter and 241 mg/cm^2 thick). They obtained an effective cross-section of $\sigma_{\mathrm{eff}} = 26 \pm 9\ \mu$b, this being an average of 20 experimental runs.

Gardner et al. [85Ga], [87Ga] investigated the absolute dipole-strength functions in the mass region of the lutetium and reported a well-tested set of 291 modelled levels for ^{176}Lu. The cross-section (in barns) for the formation of some J^{π} states, at an energy E_{γ} above the ground state, $I^{\pi 0}$, is given by

$$\sigma_{\gamma a}^{J\pi}\left(E_{\gamma}\right) = 3.85 \times 10^{3}\left(\frac{2J+1}{2I+1}\right)\sum_{\chi l} E_{\gamma -}^{2l+1}\ f_{\chi l}\ \left(E_{\gamma}\right) \tag{74}$$

where the sum over multipole type χl, is properly taken to ensure the conversion of angular momentum and parity. The total γ-ray absorption cross-section, $\sigma_{\gamma a}(E_{\gamma})$, is obtained by summing over all the J^{π} excited states so produced. The conversion makes use of the relation $T_{\gamma\chi 1}(E_{\gamma}) = 2\pi E_{\gamma}^{2l+1} f_{\chi l}(E_{\gamma})$. The calculation for photon transitions of sufficient energy to reach allowed states in the continuum proceeds by calculating the cross-section of such J^{π} state each; then the γ-ray cascade is followed from each initial state of this type until it terminates at either the ground state or the isomer. For low-energy incident photons that reach energy bins occupied solely by discrete levels, a determination is first made of the available spins and parities associated with the levels in the bin, and then the allowed cross-section is evenly distributed over those levels with the appropriate J^{π} values. If a multipole transition can reach J^{π} states that are not presented by the levels in the bin, that portion of photon absorption cross-section is deleted. Along with this portion, the cross-section places an upper limit on the photon absorption process. The principal approximation in their calculation was that the K quantum number was not considered when the transition was from one discrete target state to an energy bin below the continuum containing a number of discrete levels.

The situation is shown in *Fig. 61.* The authors [87Ga] made several features: (1) The levels are spaced several keV apart, and a line source of ^{60}Co γ-rays has a good chance of being "off-resonance", with energies between levels. (2) Only a small subset of the excited levels in ^{176}Lu can be reached by dipole transition from the 7^{-} ground state. The $E2$ transition probabilities are so small at these energies that they contribute little to the isomer production. (3) It often happens that the states reached by the transitions of a given multipole type

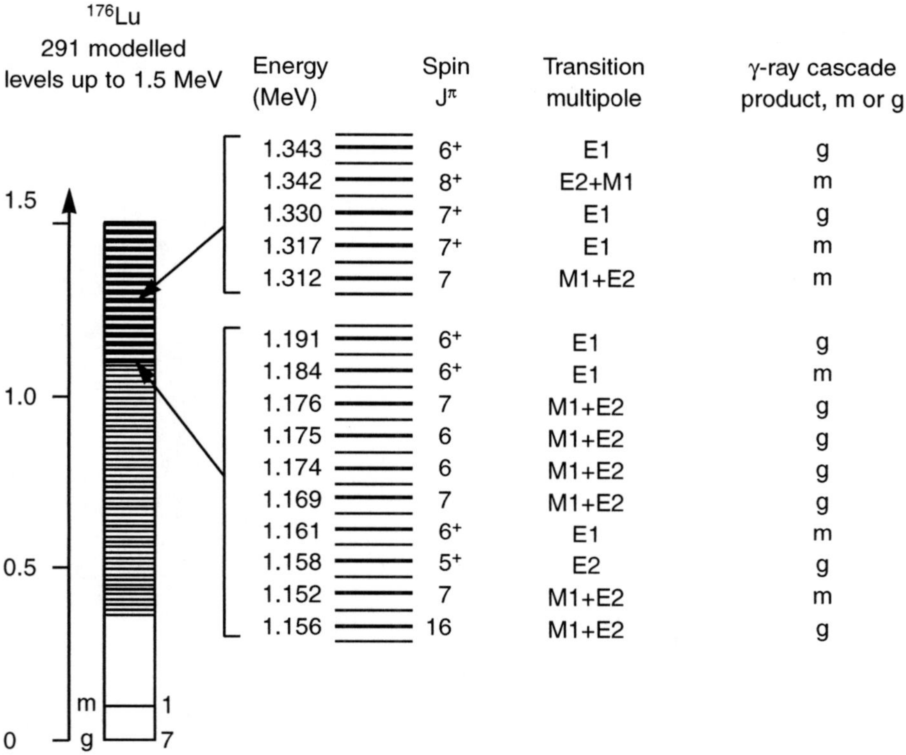

Fig. 61. Modelled discrete nuclear levels in ^{176}Lu. The regions around 1.17 and 1.33 MeV, and the ^{60}Co primary γ-ray energies are shown in detail. The data are taken from [87Ga]

do not cascade down to the same final level, be it isomer or ground state.

It is, therefore, most likely that photo-production from a discrete-energy γ-ray source will yield only some lower limit of the photoexcitation cross-section; this lower limit is more a measure of the spectrum emitted by the source itself than a property of the target. It is also believed that a better experiment would be to irradiate ^{176}Lu with a *bremsstrahlung* source or some other source whose photon energies are spread over at least 50 to 100 keV.

Klay et al. [91K1] studied the structure of ^{176}Lu and its astrophysical consequences. The 5^- state at 838.6 keV (which belongs to the $K^\pi = 4^-$ band) acts as the main mediating level, since it connects the two states by allowed transition *(Fig. 62)*. If K^- forbidden transition also contributes, the 437 keV member of the 0^+ bound may be important as well.

The cross-sections and other values may be found in *Table 23* with the results also given in the higher energy range. Carroll et al. [91Ca2] examined the isomers of 19 nuclides. Four accelerators were used as sources of *brems-*

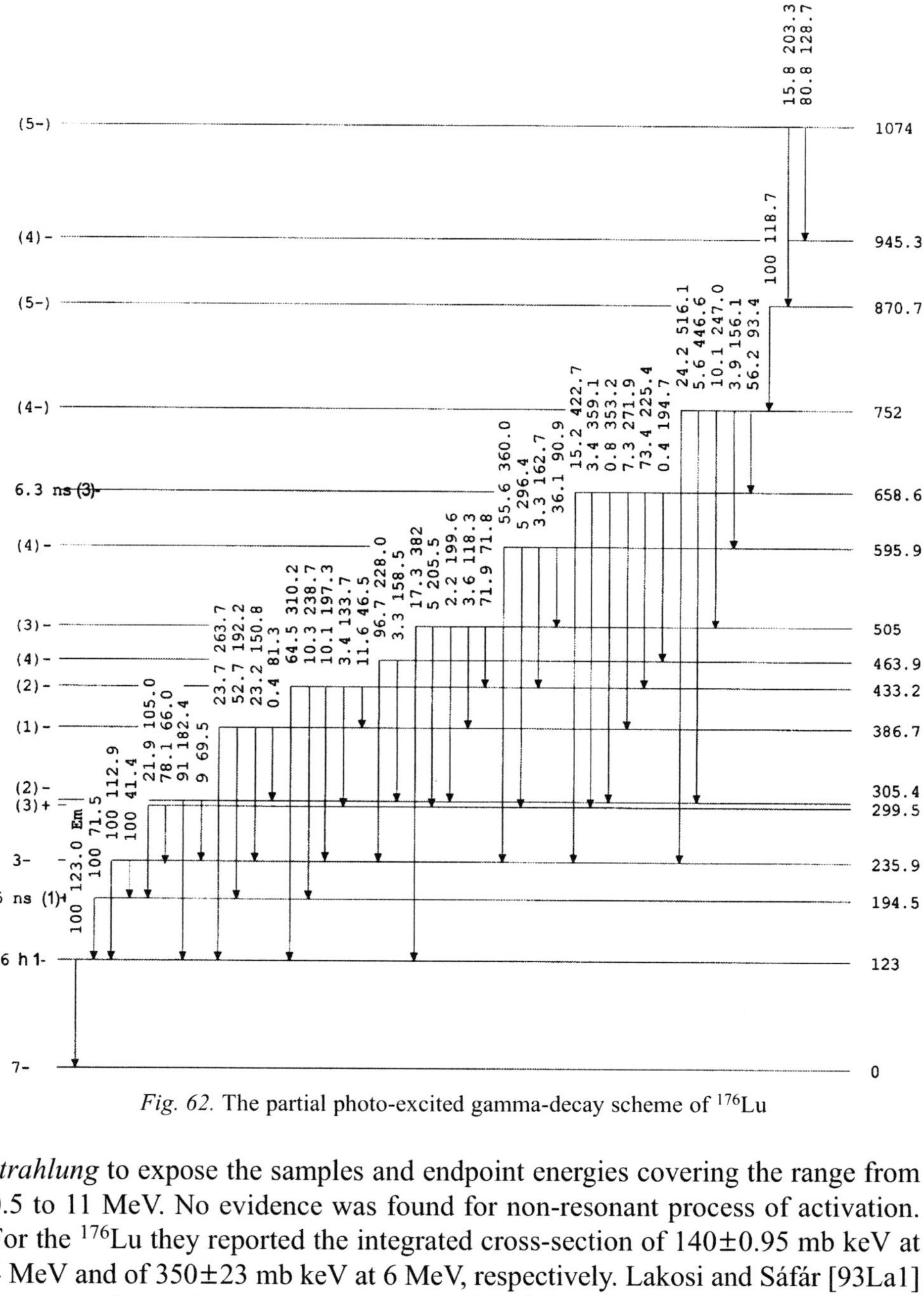

Fig. 62. The partial photo-excited gamma-decay scheme of ^{176}Lu

strahlung to expose the samples and endpoint energies covering the range from 0.5 to 11 MeV. No evidence was found for non-resonant process of activation. For the ^{176}Lu they reported the integrated cross-section of 140±0.95 mb keV at 4 MeV and of 350±23 mb keV at 6 MeV, respectively. Lakosi and Sáfár [93La1] estimated the excitation of isomers around and above neutron binding energy by (γ, γ') reaction, and a saturation value of 5.2 mb MeV was obtained. Lakosi

[96La] excited a series of isomers using *bremsstrahlung* from a 4 MeV, 25 μA linear accelerator; and reported integrated cross-section of 310 μb MeV at 4 MeV endpoint energy, and effective cross-section of 46 μb for flux >1 MeV, respectively.

6.19.2. Stellar clock or thermometer?

The ^{176}Lu nucleus also displays interesting behaviour beyond the standpoint of nuclear physics in stellar environments. This particular isotope is made only by the *s*-process of stellar nucleosynthesis, being shielded from *r*-process synthesis by the stable isotope of ^{176}Yb. With a ground state half-life of 3.5×10^{10} y, ^{176}Lu would function as an excellent galactic *s*-process chronometer were it not for the 3.7 hours isomeric state at 123 keV; this β-decays completely to ^{176}Hf. Numerous investigators have addressed (with various degrees of detail) the problems associated with the use of ^{176}Lu in understanding the *s*-process of nucleosynthesis; these researchers include Allen, Beer, Gardner, Käppeler, Kellogg, Norman, Ward [81Al], [81,82Be1,2], [77,80Wa], [84No], [88Ga3], [91Zh]. As illustrated in the paper of Beer [82Be1,2] and Klay et al. [91Kl] (see *Fig. 63*), the isomeric population leads to a significant branching in the *s*-process. In ^{176}Lu, which is shielded by ^{176}Yb from contributions the *r*-process, the nuclei are partly synthesised as ^{176}Lu^{m} (half-life = 3.68 hours) and partly as ^{176}Lu (3.6×10^{10} years).

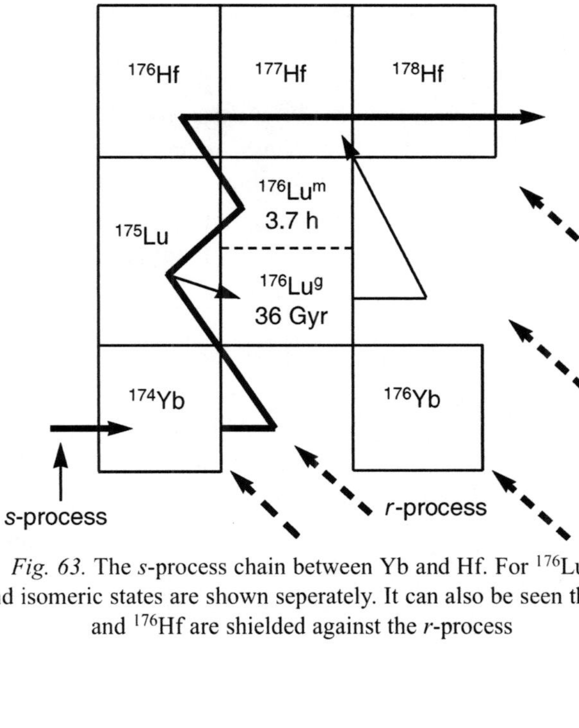

Fig. 63. The *s*-process chain between Yb and Hf. For ^{176}Lu, ground and isomeric states are shown seperately. It can also be seen that the ^{176}Lu and ^{176}Hf are shielded against the *r*-process

The fraction of material generated as $^{176}Lu^g$ is designated by f_n. For the average age T of the s-process, $R = N^*(^{176}Lu/N_{\odot}(^{176}Lu) = \exp(\lambda T)$, where λ is the beta decay rate of ^{176}Lu, and N^* (^{176}Lu) and $N_{\odot}(^{176}Lu)$ are the original and solar system abundances of ^{176}Lu, respectively ($N_{\odot}$ being related to the time of solidification of solar system). The branching factor is normally connected to the population probability P of the isomeric state in ^{176}Lu, obtained via neutron capture on ^{175}Lu. Therefore $f_n = 1 - P = 1 - \sigma^m(^{175}Lu)/\sigma(^{175}Lu)$ can be written where $\sigma^m(^{175}Lu)$ and $\sigma(^{175}Lu)$ are the Maxwellian-averaged capture cross-section to the isomeric state of ^{176}Lu and the total capture cross-section of ^{175}Lu, respectively. A measurement of these cross-sections makes it possible to determine f_n. The average age T of the s-process can then be calculated, if the other remaining quantities are also determined experimentally. It was stressed by Ward [76,77Wa] and Beer [81Be2] that f_n might be altered by thermal equilibration effect in the hot stellar environment where the s-process nucleosynthesis takes place. In conclusion , it must be pointed out that the s-process age can be calculated solely from measured capture cross-section and solar abundance ratios. The only assumption about the s-process is the well-accepted fact that, between closed shells, the product σN is a smooth, nearly constant function of A. But on the basis of the improved level scheme of ^{176}Lu [91Kl], [91Le], it was demonstrated that the stellar s-process production, as well as the stellar beta-decay rate of that nucleus, depend strongly on temperature.

The main conclusion from the above-mentioned studies is that, as a consequence of the coupling between ground state and isomer, the s-process production of the ^{176}Lu becomes a sensitive function of temperature. It was found that its original abundance is determined by the neutron densities and temperatures at the s-process site; and remains, therefore, too uncertain for using ^{176}Lu as a chronometer for the age of the s-process. Instead, it can be considered as a suitable s-process thermometer, yielding a temperature range between 2.4×10^8 and 3.6×10^8 K. This is in good agreement with similar estimates from other s-process systems of branching and with current stellar models.

6.20. $^{177,178,179,180}Hf$

There are six natural isotopes of the element hafnium. Four of these have isomer states with different spins, energies and half-lives: ^{177}Hf ($J_0 = 7/2^-$; natural abundance = 18.6%), ^{178}Hf ($J_0 = 0^+$; natural abundance = 27.1%), ^{179}Hf ($J_0 = 9/2^+$; natural abundance = 13.74%), and ^{180}Hf ($J_0 = 0^+$; natural abundance = 35.2%).

6.20.1. Photoexcitation data

The most characteristic data of these isomers are:

– ^{177}Hf: $J_{m1} = 23/2^+$, $E_{m1} = 1312.4$ keV and $t_{m1} = 1.1$ s; $J_{m2} = 37/2^-$, $E_{m2} = 2739.9$ keV, and $t_{m2} = 51.4$ min.
– ^{178}Hf: $J_{m1} = 8^-$, $E_{m1} = 1147.4$ keV, and $t_{m1} = 4$ s; $J_{m2} = 16^+$, $E_{m2} = 2446$ keV, and $t_{m2} = 31$ y.
– ^{179}Hf: $J_{m1} = 1/2^-$, $E_{m1} = 375$ keV, and $t_{m1} = 18.67$ s; $J_{m2} = 25/2^-$, $E_{m2} = 1105.84$ keV, and $t_{m2} = 25.5$ d.
– ^{180}Hf: $J_m = 8^-$, $E_m = 1145.5$ keV, and $t_m = 5.5$ h.

In the literature there is no mention of photoexcitation and the isomers of the ^{177}Hf nucleus. For this reason it may be assumed that the spin-difference values between the ground state and activation level are too high. In other words, there appears to be a strong hindrance with a (γ, γ') reaction to produce both of the isomeric states of ^{177}Hf. In the energy range above neutron-binding energy (10–30 MeV) excitation of these isomer states may happen with multiple-step transitions with a (γ, γ') reaction, but the appearance of a competitive reaction could be problematic in the evaluation of measured data.

The energy diagram of ^{178}Hf with the isomer decays and the separation energies of the proton and neutron is shown in *Fig. 64.* Those for ^{180}Hf [71Gu] are presented in the next section, together with those of ^{180}Ta [84Es], [86Ke], [86Le], [97La].

There were a few attempts to produce the isomers of ^{178}Hf and ^{180}Hf. Kaminiski and Kojima [63Ka] irradiated $^{178,\,180}$Hf and other even-even nuclides with a linear accelerator (4–6 MeV, average beam current of 20–100 μA) and they observed no isomeric activity.

Abrams and Pelekis [66Ab] irradiated a hafnium target with the mono-energetic γ-rays of a ^{116}In^{m} radioactive source ($E_{\mathrm{max}} = 2112$ keV) produced in a nuclear reactor loop; they did not observe any activity. Veres et al. [73Ve2] investigated the possibilities of photoexcitation in the respective cases of five (^{90}Zr, ^{178}Hf, ^{180}Hf ^{190}Os, and ^{204}Pb) even-even nuclei by 1.38 and 2.75 MeV γ-rays from a ^{24}Na (92.5 TBq) source. No measurable isomeric activity was obtained for any of the samples under investigation. Therefore the only useful information furnished by these experiments was that the upper limits of the activation cross-section per primary γ-quantum could be estimated as 10 pb for ^{178,180}Hf. The conclusions are very similar to even-even nuclei earlier mentioned.

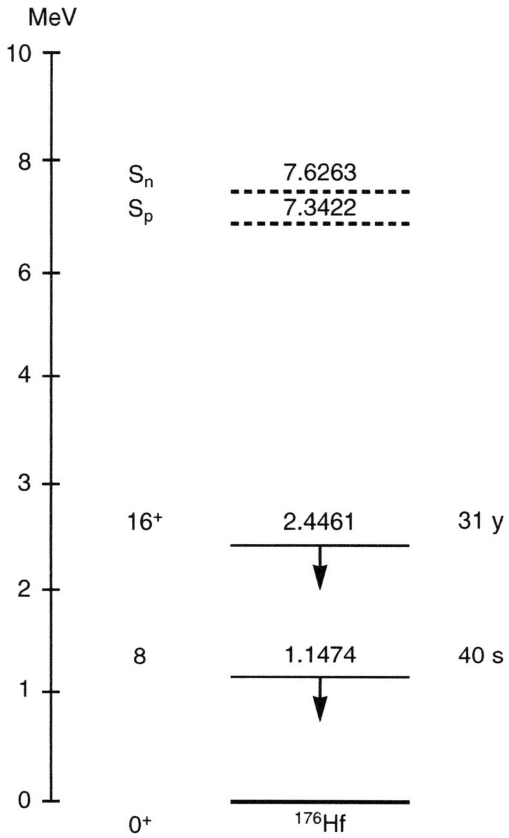

Fig. 64. The decay diagram of ^{178}Hf with the two isomer states and the proton and neutron separation energies

However, Gangrsky et al. [96Ga] presented a successful experiment on the possibilities of the isoneric state excitation of ^{180}Hf by (γ, γ') reaction. They measured the excitation of ^{180}Hf as a function of energy (between 5–15 MeV). They observed a value of 209±44 nb MeV for the integrated cross-section for saturation.

The photoexcitation of ^{179}Hf most probably occurs because it is a single particle excitation with low spin and energy in the isomeric state. Several attempts have been made to produce the isomeric state of ^{179}Hf using strong ^{60}Co and ^{116}In^{m} γ-ray sources. Veres [73Ve2] irradiated a natural hafnium target with γ-rays of ^{60}Co, having intensity of 48.5 TBq. From the measured isomeric activity (150±34 counts per minute) he derived an effective cross-section of 10 nb. The integral cross-section, supposing an activation level of about 1.1 MeV, is about 60 mb eV *(Fig. 65)*. Zaparov et al. [73Za] reported a smaller value: ≤2.1

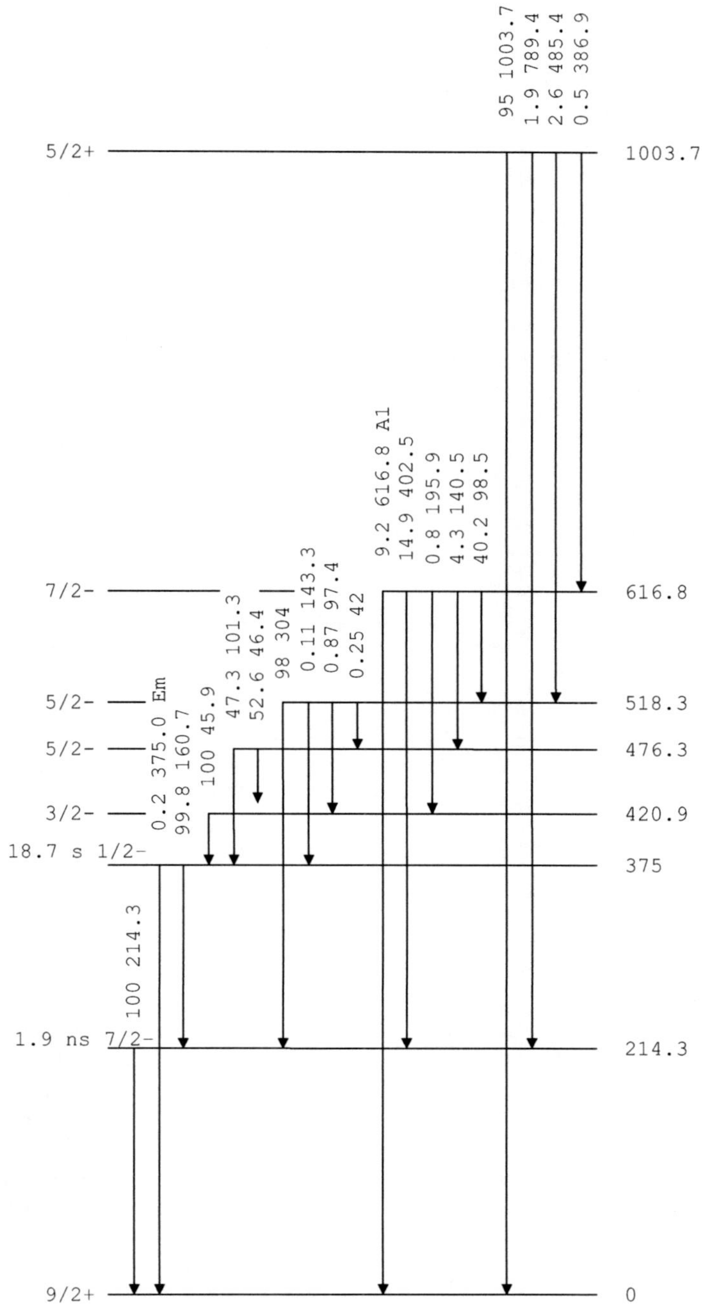

Fig. 65. A partial scheme of the γ-excitation and γ-decay of ^{179}Hf

mb eV in the 1.17–1.33 MeV energy range. Abrams and Lakosi [69Ab], using $^{116}In^{m}$ and ^{24}Na radioactive sources, reported 2±0.5 μb and 29±6 μb for the effective cross-sections, respectively.

Several photoexcitation experiments have also been carried out with *bremsstrahlung* of linear accelerators. The investigated energy ranges were from 1.5 MeV to 15 MeV. Carroll et al. [91Ca2] and [92Co] studied the isomer yields of ^{179}Hf in the 0.5–11 MeV energy range. Mazur et al. [93Ma1] measured the cross-section of ^{179}Hf at 4–15 MeV energies in ΔE = 0.5 MeV steps. For a (γ, γ') reaction below the (γ, n) threshold, the energy dependence of the energy ratios σ_m/σ_{tot} was considered. Lakosi and Sáfár [93La1] produced semi-empirical estimated values for the saturated integrated cross-sections of 31 isomers, among them ^{179}Hf, too. The authors measured the effective and integrated cross-sections of ^{179}Hf [95La2] by 4 MeV endpoint energy *bremsstrahlung* (data in *Table 25*).

Table 25. The excitation cross-sections of the ^{179}Hf nucleus

E_{exc} (MeV)	σ_{eff} (nb)	σ_{int} (mb eV)	Σ_{int} (mb keV)	Refs
~1.1	10±2.4	60±14		[73Ve2]
<2.112	2±0.5			[66Ab]
<2.75	29±6			[69Ab]
1.17–1.33		≤2.2		[73Za]
4			94±1.2	[91Ca2]
6			250±11	
1.5±0.2			0.4±0.1	[92Co]
2.5±0.1			12±2	
3.0±0.15			60±10	
4.3±0.2			750±150	
7–9			4600 a)	[93La1]
8			4400±500	[93Ma1]
10			4800±800	
4	89000		610	[95La2]

a) calculated value

6.20.2. Applicability of Hf isomers for a graser pump

With respect to hafnium isotopes with isomeric states, here consideration is given to the possibility of whether hafnium can be used in a graser pump. In *Table 9* the isomers of stable nuclei are identified. These store exceptionally high densities of energy. It can be seen in column 8 of *Table 9* that the energy storage for the best of the isomers, $^{178}Hf^{m2}$, is more than a gigajoule per gram. Since isomers derive their long half-lives from their poor coupling to an electromagnetic field, it was traditionally thought to be impossible to trigger the release of the stored energy. However, Collins and Carroll [97Co2] showed the existence of a giant pumping resonance at an energy of about 2.5 MeV in nuclei with masses near 180. In effect, this resonance provided a "gateway" (activation) state through which the selection rules making an isomer may be violated. The absorption of an X-ray photon of that energy excites the system to such a level which is very strongly coupled to the electromagnetic fields. The sum of the stored energy and that of trigger X-ray photon would then be promptly emitted, or dumped, as gamma rays.

The concept behind dumping the energy from controlled fractions of isomeric populations has been demonstrated in a series of experiments with the materials of $^{177}Hf^{m}$, $^{178}Hf^{m2}$, $^{179}Hf^{m}$ and $^{180}Ta^{m}$, at the University of Texas at Dallas (UTD) and at the *Institut für Kernphysik*, Darmstadt. The triggered release of stored energy in electromagnetic waves was reported to be promising.

In the 1990s a research programme was started as a result of the JINR-Dubna, CSNSM and IPN Orsay co-operation and it has become a wide programme, involving 16 institutes from six countries [97Og]. The aim of the programme has been the separation of the $^{178}Hf^{m2}$ isomer produced, and its utilisation in various different physical experiments. The first step towards this was isomer production in microweight quantities, and the second step involved its purification.

Several different reactions have also been used with the aim of producing $^{178}Hf^{m2}$ for spectroscopic studies. The used reactions, irradiation conditions and the isomeric/ground ratio are given in *Table 26.* The most convenient $^{178}Hf^{m}$ production is by means of the $^{176}Yb(\alpha, 2n)^{178}Hf^{m2}$ reaction.

The optimum energetic interval lies from 28 to 36 MeV, where the absolute cross-section of the isomer formation is at a maximum. For $^{178}Hf^{m2}$

Table 26. Production of the $^{178}Hf^{m2}$ isomer

Reaction	E_{pr} (MeV)	σ_{m2}/σ_g
^{177}Hf (n, γ)	Thermal	10^{-9}
^{181}Ta (γ, p2n)	<85	low
^{181}Ta (α, αp2n)	120	–
^{181}Ta (p, α)	92.5	10^{-3}
^{176}Yb (α, 2n)	26	0.01
	27	0.005
	32	0.05
	40	0.08

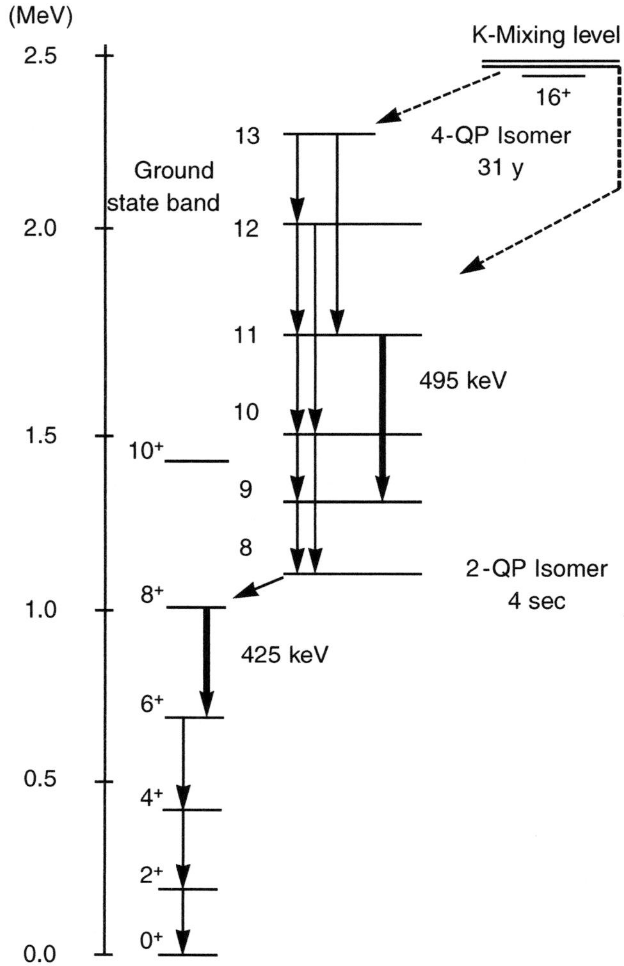

Fig. 66. Energy level diagram for ^{178}Hf, showing a selection of levels relevant to the experiment of [99Co]. The spontaneous decay of the 31 y 16^{+} isomer is shown by the thin arrows. The thick solid arrows show components observed in the experiment from the particular cascade from the decay forced by the X-ray irradiation

isomer-production in a quantity of hundreds of nanograms, the cyclotron U-200 of the Flerov Laboratory of Nuclear Reactions has been modified for intensive α-particle-beam mode operation. As a result, up to 150 μA of $^{4}He^{2+}$ ions can be obtained on the target with 36 MeV energy. The effective beam time needed in order to produce about 5×10^{14} $^{178}Hf^{m2}$ atoms, using a 100 μA current of 36 MeV He ions, amounts to 500 hours.

In recent years [99Co] seven institutes have carried out experiments for accelerated emission γ-rays from the $^{178}Hf^{m2}$ isomer induced by X-ray irradiation. A sample of 6.3×10^{14} nuclei of the 4-quasi-particle isomer of ^{178}Hf, having a half-life of 31 years and an excitation energy of 2.446 MeV, was irradiated with X-ray pulses from a device; the latter was operated at 15 μA to produce *bremsstrahlung* radiation with an endpoint energy set to 70 or 90 keV. The spectra of the isomeric target were taken with a high purity Ge detector. Intensities of selected transitions in the normal decay cascade of the ^{178}Hf isomer was reported to be increased by about 4%. Such an accelerated decay is consistent with an integrated cross-section of 1×10^{-21} cm^2 keV for the resonant absorption of X-rays and for the induction of γ-decay. However, this result is widely disputed.

The relevant energy levels of the ^{178}Hf nuclide are shown in *Fig. 66.* The spontaneous decay of the 31-yr isomer – shown by the thin arrows – is entirely through a $K^\pi = 8^-$ band, followed by a cascade through the ground state band (GSB) beginning with the 8^- member. Particular attention was focused upon the $(8^+ \rightarrow 6^+)$ and $(11^- \rightarrow 9^-)$ lines at 426.4 and 495.0 keV, respectively (shown by thick arrows). The 495 keV line showed the greatest enhancement when the target was irradiated with *bremsstrahlung*. The absence of the 574.2 keV $(13^- \rightarrow 11^-)$ transition was also noted; this normally feeds the 495.0 keV transition during the spontaneous decay of the ^{178}Hf isomer, as seen by the thin arrows in *Fig. 66.*

The authors reported in their paper [99Co] that the resonant absorption of an X-ray photon, with energy in the order of 40 keV, can induce the prompt release of the 2.446 MeV energy stored by the isomer into freely radiating states. This is an energy gain of about 60 keV. Only 9 mW cm^{-2} of the total X-ray power in *bremsstrahlung* was required at the target. Further research is needed to prove the result referred to above.

6.21. $^{180}Ta^m$

The nuclide ^{180}Ta bears the distinction of being nature's rarest quasi-stable isotope (its natural abundance is 0.0012%). Though the short-lived (eight-hour) radioactivity of $^{180}Ta^g$ was identified in 1938 by Oldenberg [38Ol], and Brown et al. [51Br], the long-lived, high spin isomer remained undiscovered until 1955 (White, Collins, and Rourke [55Wh]). However, it was first demonstrated in the years 1977–80 [77Wa], [79Wa1], and [80Sh] that the naturally-occurring isotope is not in its ground state but is in fact an isomer at an excitation energy of 73 keV and with a spin parity of $J^\pi = 9^-$. An impressive lower limit has been established for the half-life of the isomeric state $t_{1/2} > 1.1 \times 10^{15}$ y. This stability is principally due to the collective behaviour of the rare-earth nuclei. The well-known 8.1 hour activity is actually of the ground state of ^{180}Ta having $J^{\pi} = 1^{+}$ *(Fig. 67)*. It decays to the ground and first-excited states of ^{180}Hf and ^{180}W, respectively [80Ry], [81No], [83De], [83Yo], [85Cu], [92Ke] [94Sc], and [97Bi].

6.21.1. Photoexcitation data of ^{180}Ta^{m}

Law and Iddings [69La] irradiated a natural tantalum target of 24 grams and a 4"×4" sheet, with ^{60}Co γ-sources and a flux of $(1.6-4)\times10^{11}$ photons per cm^2 second.They did not observe any activity of ^{180}Ta *(Table 27)*. Norman et al. [84No] also investigated the production of ^{180}Ta^{g}. According to Carroll et al., from these results it follows that either this isomer must have been singularly stable against photonuclear de-excitation – ^{180}Ta^{m}(γ, γ')^{180}Ta^{g} – at the time of creation, or the corresponding temperatures must have been too low to produce photons capable of pumping such a reaction [89Ca]. Bikit at al. [99Bi1,2] observed the population of the ground state of ^{180}Ta using a strong (2.78 PBq ~75 kCi) ^{60}Co source for the irradiation of a natural tantalum target of 8.647 grams. They used a low-background iron shield detector for the measurement of the ground-state activity and they determined 0.12±0.002 μbarns for the effective cross-section and 0.13±0.04 μbarns MeV for the integrated cross-section.

Collins et al. [90Co], investigated the excitation function of the ^{180}Ta^{m}(γ, γ')^{180}Ta^{g} reaction in the 2–5 MeV energy range. In their paper the authors drew three conclusions: (a) there are no significant activation levels below 2.8 MeV; (b) there are activation levels at 2.8 and 3.6 MeV excitation energy with integrated cross-sections of 1.2×10^{-22} and 3.5×10^{-22} cm^2 eV, respectively; (c) these activation levels lie high enough to allow ^{180}Ta^{m} to survive under *s*-process circumstances independently of the cross-sections. Németh [92Né1] commented on the resonant excitation of ^{180}Ta, based on his experiments. He established an upper limit of 14 nb for the overall cross-section at 1.33 MeV and deduced a value of 0.52(20) mb at 4 MeV. Németh's observations raised an issue with regard to the calculated photon flux used by Collins et al. Namely, it should be rechecked with respect to its connection with the observation of intermediate states in the resonant (γ, γ') reaction at 2.8 and 3.6 MeV, with large integrated cross-sections of 120 and 350 mb keV, respectively. Carroll at al. [92Ca], in reply to Németh [92Né1], presented independent experimental proof that both the total flux and the spectral distribution of the *bremsstrahlung* are well understood. The indirect evidence against the integrated cross-section magnitudes, as put forward by Németh, was shown to be unfounded.

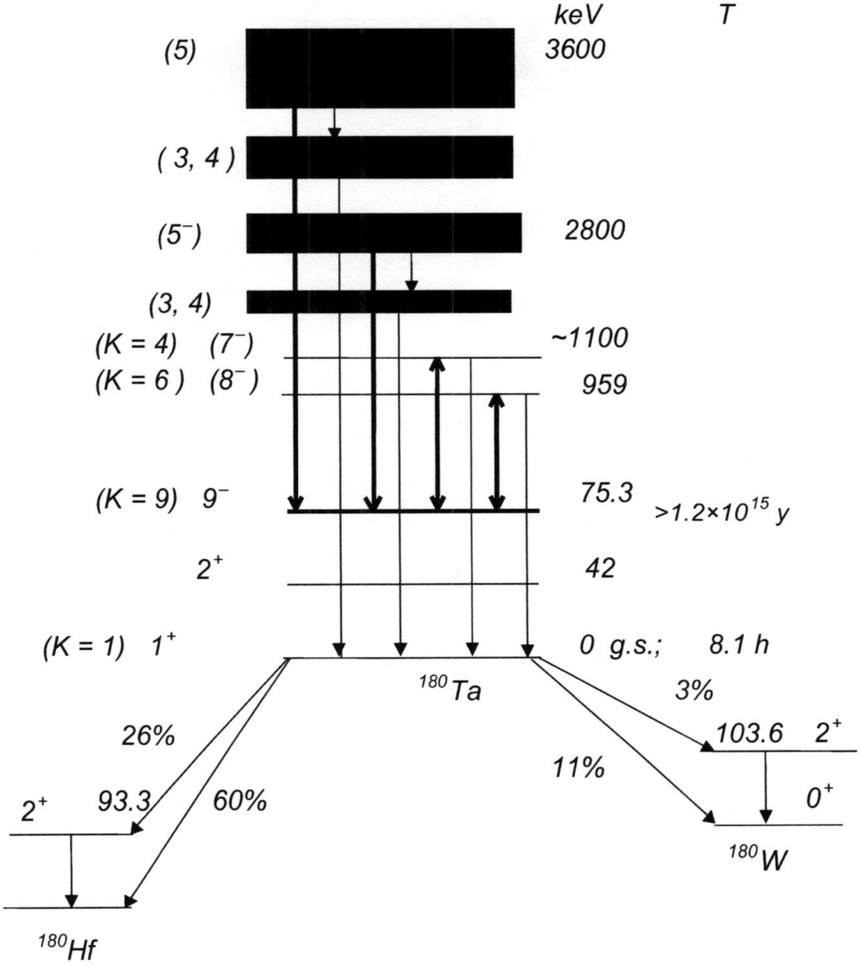

Fig. 67. A possible photoexcitation and decay scheme of ^{180}Ta^{g}.
The data are derived from only a small number of the experiments which have been carried out

Absolute methods developed for calibration of *bremsstrahlung* spectra by von Neumann–Cossel et al. [94Neu] and Schlegel et al. [94Sc] seem to confirm the result given by Collins. However, a later Coulomb-activation experiment performed by Schumann et al. [97Sc] appeared not to confirm this finding, while the preliminary analysis of the in-beam experiment indicates Coulomb-excited levels in ^{180}Ta around 700 and 1050 keV [97Neu].

According to Bikit et al. [99Bi2], the low irradiation intensity measured by Norman et al. [84No], as well as the insufficient shielding and especially the

Table 27. The excitation cross-sections for photoproduction of $^{180}Ta^g$

Irradiation facility and/or intensity	Excitation energy (MeV)	σ_{int} (mb eV)	Σ_{int} (μb MeV)	$\sigma_{experiment}$ (nb)	References, remarks
^{60}Co, 0.2 ; 1.1 PBq	<1.33			no isomer activity	[69La]
^{60}Co				no isomer activity	[84No]
^{60}Co	<1.33			<14	[92Né1]
^{60}Co, 2.78 PBq	~1.1 or 0.959	130±40		120±20	[99Bi2]
LINAC, 6 MeV	2.8		120±20		
	3.6		350±50		[90Co]
Four accelerators,	4		180±66		
0.5–11 MeV	6		350±23		[91Ca2]
LINAC, 4 MeV	4		300±110	32000±11000	[99Bi1]

employment of lead shielding [92Né2] (where the fluorescence of lead X-rays disturbs the detection in the energy region below 72 keV), might explain why these experiments – ([92Né2] and [84No]) – failed to detect the depopulation of ^{180}Ta through an excitation level of E<1.33 MeV.

Bikit et al. [99Bi1] reinvestigated the photoexcitation of ^{180}Ta with a LINAC-produced 4 MeV endpoint energy *bremsstrahlung*. The detected values of the effective and integrated cross-sections are σ_{eff} = 32±11 μb and σ_{int} = 300±110 μb MeV, respectively. The authors concluded that their results correspond with those of Collins et al. [90Co].

What remains open is an understanding of the microscopic nuclear structure underlying the breakdown of K conservation and the strong coupling with the intermediate states. Further experiments are clearly warranted.

6.21.2. Nucleosynthesis of $^{180}Ta^m$

The production (nuclear astrophysical) and the decay (laboratory) schemes of ^{180}Hf in the mass-180 chain were presented by Beer and Ward [81Be1], and the isomer-to-isomer decay of ^{180}Hf and the nucleosynthesis of $^{180}Ta^m$ were studied by Kellogg and Norman [92Ke]. The nuclear level structure along the mass-180 isobaric chain of significance to the stellar nucleosynthesis of $^{180}Ta^m$ is shown in *Fig. 68*. The most important quantities are the fractional branching into the $^{180}Hf^m$ (5.5 hours) through both the 30 keV neutron capture by ^{179}Hf (s-process) and by the β-decay of ^{180}Lu (r-process). These are then followed by the possible β-decay branching from the $^{180}Hf^m$ into the $^{180}Ta^m$. The data are only a small representation of a great deal of experimental work.

The excitation of the ground state of the mass-180 isotope of tantalum with a $^{180}Ta^{m}(\gamma, \gamma')^{180}Ta^{g}$ reaction is very important in nucleosynthesis. The $(\gamma, \gamma')^{m}$ experiments show that the effective half-life in a stellar environment might be seriously affected by destructive photoexcitation as a consequence of thermal coupling between the isomer and the ground state. If the first activation level lies below or at 1 MeV the $^{180}Ta^{m}$ cannot survive at typical $(2.9-3.3)\times 10^{8}$ *K* *s*-process temperature [90Kä] because of the very short residence time. To locate the first activation level more precise spectroscopic data are needed.

Norman et al. investigated some effects of high temperature and density on the half-life of $^{180}Ta^{g}$. If it were possible for $^{180}Ta^{g}$ and $^{180}Ta^{m}$ to come into thermal equilibrium, then their abundance ratio would be given by the expression:

$$\frac{N_m}{N_g} = \frac{2J_m + 1}{2J_g + 1} \exp\left(-\frac{E_m}{kT}\right), \qquad (75)$$

where N_g and N_m are the abundances and J_g and J_m are the spins of $^{180}Ta^{g}$ and $^{180}Ta^{m}$, respectively, E_m is the excitation energy of $^{180}Ta^{m}$, k is Boltzmann's constant, and T is the temperature. Inserting the experimentally-determined values for these parameters yields $N_m/N_g = 6.33 \exp(-9.3/T_8) = 0.45$, at the canonical *s*-process temperature of $T_8 = 3.5$ ($kT = 30$ keV), for example (where T_8 is the temperature in units of 10^8 *K*). Such an equilibrium between $^{180}Ta^{g}$ and $^{180}Ta^{m}$ would have serious consequences for the nucleosynthesis of ^{180}Ta, for it would imply that the "effective" lifetime of $^{180}Ta^{m}$ under stellar conditions is much shorter than its laboratory value. In the laboratory, $^{180}Ta^{g}$ decays via electron capture approximately for 80% of the time, as can be seen in *Fig. 68.*

However, under stellar conditions of high temperature and density, the number of bound atomic electron can be substantially reduced thereby increasing the half-life of $^{180}Ta^{g}$. Norman et al. [84No] calculated the half-life of the ground state of tantalum and taking $T_8 = 3.5$ and density 10^3 g cm^{-3} as an example, the half-life is 34.2 hours. If the $^{180}Ta^{g}$ and $^{180}Ta^{m}$ are in equilibrium under these conditions, the effective half-life of the nucleus becomes 49.5 h. Thus if $^{180}Ta^{g}$ and $^{180}Ta^{m}$ can equilibrate, the net amount of long-lived $^{180}Ta^{m}$ produced in stars could be substantially reduced.

Depopulation of the isomer was observed with large yields in photoexcitation experiments using *bremsstrahlung* (Németh et al. [92Né2], Carrroll et al. [89Ca], Collins et al. [90Co]) but the intermediate state energies (2.8 and 3.6 MeV) were too high to play a role here. Attempts [84No, 92Né2] to excite states at low energies <1,33 MeV) with intense ^{60}Co sources showed no signals. There was neither proven nor disproven a synthesis of ^{180}Ta in the *s*-process. Belic et al. [99Be] investigated the depopulation of the quasi-stable isomer in

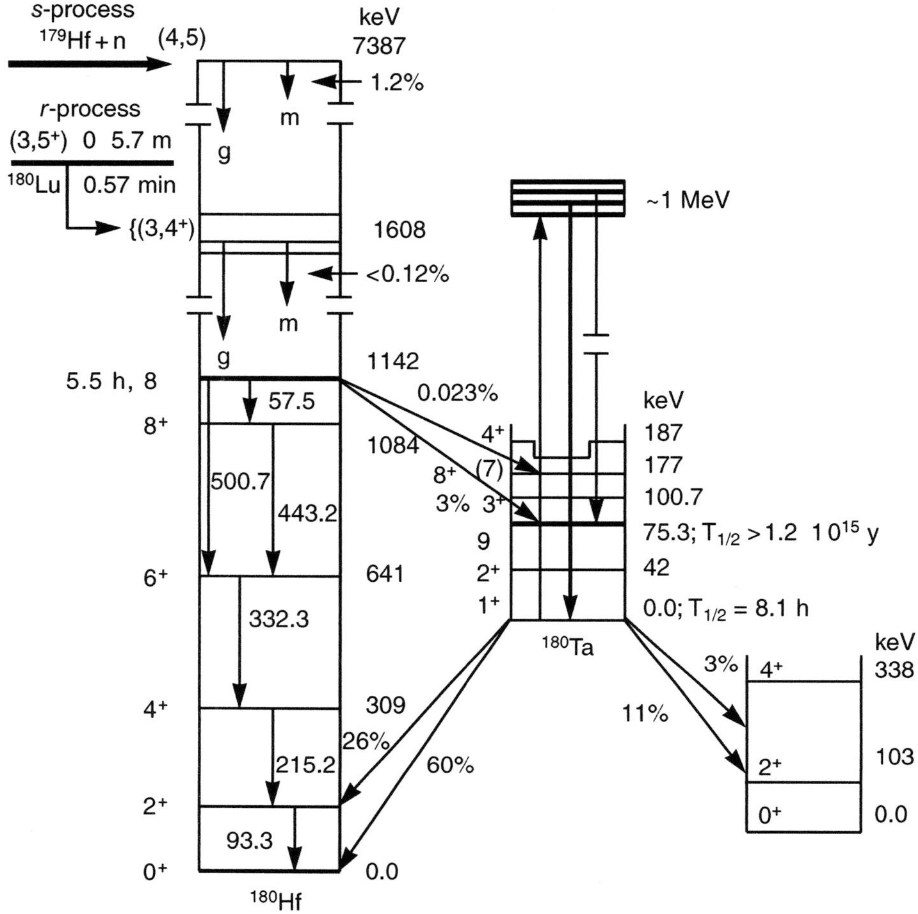

Fig. 68. The nuclear level structure along the significant isobaric chain of stellar nucleosynthesis of ^{180}Ta

^{180}Ta (9^-, at 75.3 keV) by resonant absorption with intense *bremsstrahlung*. They demonstrated the depopulation of the long-lived isomer to the short-lived ground state by resonant absorption of photons through a possible intermediate state with energy as low as 1.01 MeV.

Among the possible mechanisms for de-exciting $^{180}Ta^m$ is photoexcitation. This process undoubtedly involves transitions from $^{180}Ta^m$ to some higher lying states of intermediate spin which subsequently decays to $^{180}Ta^g$, as it can be seen in *Fig. 68.*

Bikit et al. [99Bi1] irradiated a natural tantalum target with 4 MeV *bremsstrahlung* and very intensive ^{60}Co sources and their experiments provided direct evidence of the $^{180}Ta^{m}(\gamma, \gamma')^{180}Ta^{g}$ depopulation below 1.33 MeV excitation energy. These experiments show that the $^{180}Ta^{g}$ effective half-life in a stellar environment might be seriously effected by destructive photoexcitation as a consequence of thermal coupling between the isomer and ground states.

6.21.3. Possibilities for pumping a graser with $^{180}Ta^{m}$

Collins and Carroll [97Co1] delivered an excellent lecture about "Progress in the pumping of a gamma-ray laser" at the 1st International Gamma-Ray Laser Workshop (GARALAS '95) Predeal (Romania). Among other things, they studied the applicability of the isomer of ^{180}Ta as a graser pump. In an experiment which they conducted in 1987, 1.2 mg of $^{180}Ta^{m}$ was exposed to *bremsstrahlung* and a large fluorescence yield was obtained [88Co]. This was the first time a (γ, γ') reaction had been excited from an isomeric target as a necessity for a gamma-ray laser. No more than the observation of fluorescence from a milligram-sized target proved that an unexpectedly large reaction channel had opened. In earlier experiments of this type it was usually grams of material that had been required.

6.22. ^{183}W

There are five natural isotopes of tungsten, and one of these, ^{183}W – with a natural abundance of 14.3%, and a spin-parity in the ground state $1/2^{-}$ – has an isomeric state. The energy, spin-parity, and half-life of this isomer are: $E_m = 309.5$ keV, $J^{\pi} = 11/2^{-}$, and $t_{1/2} = 5.15$ seconds, respectively.

In 1963 a comprehensive experiment was carried out for the application of a photoexcitation process with activation analysis by Kaminishi and Kojima [63Ka]. They used a 6 MeV linear electron accelerator (beam energy: 6–8 MeV; average beam current: 20–100 μA). This experiment was the first one in which the $^{183}W^{m}$ was produced with photons. They measured the induced activation rate of the isomer as being 1.2×10^{6} counts per minute, and the half-life of the isomer state was determined as having a value of 5.4 ± 0.1 seconds.

Abrams [68Ab] irradiated a natural tungsten target with the γ-rays (E_{max} = 2.112 MeV) of a $^{116}In^{m}$ radioactive radiation source produced in a neutron reactor loop. He observed weak isomer activity and reported a value of <70 picobarns for the overall cross- section of $^{183}W^{m}$. Abrams and Lakosi [69Ab] determined about <43 nanobarns using the 3.7 TBq (100 Ci) initial activity of a ^{24}Na radiation source (E_{max}= 2.75 MeV). Veres et al. [73Ve2] also irradiated a natural tungsten target with a stronger ^{24}Na source (98 TBq ~ 2650 Ci) and they observed the value of the effective cross-section to be 34 nanobarns. Mazur et al. [93Ma1] measured the cross-section of γ inelastic scattering with excitation of $^{183}W^{m}$ at 4–15 MeV energies in $\Delta E = 0.5$ MeV steps and they reported values of 0.35±0.05 mb MeV, and 0.37±0.06 mb MeV at 8 MeV and 10 MeV irradiation energy, respectively. Lakosi et al. [95La2] – using a 4 MeV, 25 μA linear accelerator – determined an effective cross-section value of 390 nanobarns for a flux below 1 MeV, and an integrated cross-section value of 2.7 microbarn-MeV at 4 MeV endpoint energy.

The main experimental data of photoexcitation are summarised in *Table 28*. In this case of the $^{183}W^{m}$ isomer there are no sufficient experimental data about the cross-section values, the energy of the activation level, the quantum numbers of the excited levels and other parameters needed to evaluate the role of the tungsten in the development of the graser.

Table 28. The excitation cross-sections of the ^{183}W nucleus

E_{exc} (MeV)	σ_{eff} (nb)	σ_{int} (mb eV)	Σ_{int} (mb MeV)	Refs
<2.112	0.07			[66Ab]
<2.75	<43			[69Ab]
<2.75	34			[73Ve2]
1.17–1.33		≤0.18		[73Za]
7–9			5	[93La1]
8			0.35±0.05	[93Ma1]
10			0.37±0.06	
<1	390			[95La2]
4 endpoint			0.0027	

6.23. Osmium isotopes

There are seven natural isotopes of osmium and out of these are three nuclei (^{189}Os, $J^{\pi}_0 = 3/2^-$, natural abundance = 16.1%; ^{190}Os, $J^{\pi}_0 = 0^+$, natural abundance = 26.4%; and ^{192}Os, $J^{\pi}_0 = 0^+$, natural abundance = 41.0%) which have isomeric states. The energy, spin-parity, and half-life of these isomeric states are: $E_m = 30.814$ keV, $J^{\pi} = 9/2^-$, and $t_{1/2} = 5.8$ hours for $^{189}Os^{m}$; $E_m = 705.4$ keV, $J^{\pi} = 10^-$, and $t_{1/2} = 9.9$ minutes for $^{190}Os^{m}$; and $E_m = 1890$ keV, $J^{\pi} = 10^-$, and $t_{1/2} = 6.1$ seconds for $^{192}Os^{m}$, respectively.

Abrams and Pelekis [66Ab] irradiated an osmium target using $^{116}In^{m}$ ($E_{max} = 2.112$ MeV) radiation sources produced in a neutron reactor loop and they observed no isomeric activity for $^{189}Os^{m}$. Law and Iddings [69La] irradiated 5 grams of OsO_4 crystals with γ-rays of 185 TBq (5 kCi) and 1.11 PBq (30 kCi) ^{60}Co sources. They did not observe any isomer activity for ^{189}Os^m and ^{190}Os^m, either.

Otozai et al. [78Ot] excited the isomeric state of ^{189}Os with 85 keV electrons. This achievement was reported as the first evidence of a new mechanism for nuclear excitation and was referred to as NEET (nuclear excitation by electron transition); NEET may occur when an electron of inner shells is forced away by bombarding electrons or X-rays, and the excitation energy is transferred from the orbital electron system to the nucleus.

The photoexcitation experiment of ^{189}Os^m was presented by Lakosi et al. [87La] at the International Symposium on Symmetries and Nuclear Structure (5–14 June 1986, Dubrovnik Yugoslavia). Natural osmium samples (in powder form) of 0.5–2.0 grams were irradiated in the radiation field of the 2.28×10^{15} Bq (61 kCi) ^{60}Co γ-ray source, and *bremsstrahlung* of a 4 MeV linear accelerator, as well as a 200 kV (15 mA) and 300 kV (5 mA) X-ray machine. The authors evaluated the photoexcitation yields, the calculated photon fluxes and the overall cross-sections for these measurements. Flat 2π and cylindrical 4π multiware proportional counters were built [92La] for counting low intensity and low energy isomeric transition lines (^{189}Os^m: 30.8 and 8.9 keV).

Based on the available data, the activation levels of 69.54, 216.68, 219.40, 275.88, 427.91, 549.91 and 599.61 keV of the ^{189}Os nucleus were identified. The proposed partial excitation and de-excitation level scheme are shown in *Fig. 69.* The results are summarised in *Table 29.*

Table 29. The excitation data for the ^{189}Os nucleus, activation level energies, and effective and integrated cross-sections

Irradiation facility	E_{exc} (keV)	σ_{eff} (μb)	σ_{int} (mb eV) Calculated;	Measured	$t_{1/2}$ (ps)	Refs
200 kV X-ray	69.5	–	$35.4B_m$	1.4	–	[95La3]
250 kV X-ray	216.7	–	$61.8/T_{1/2}$	81±9	77±9	
	219.4	–	20.1	–	–	
300 kV X-ray	275.9	–	12	–	–	
^{137}Cs , 56 TBq	Σ8 Al	1±0.2				[95La4]
^{60}Co, 3000 TBq		2.69±0.54				
4 MeV, LINAC	<4000	41.1±5.2 *		278±42 mb keV		[96La]
Calculated	7–9		11 mb MeV			[93La1]

B_m = isomer branching, $g = (2J_a+1) / (2J_g+1)$, J_a, are spins of the activation level and ground states, respectively. $t_{1/2}$ is the half-life of the activation level

$\Sigma8\ Al$ = 69.5, 216.4, 219.4, 275.9, 427.9, 531.5, 549.9, and 599.6 keV levels

* for a flux >1 MeV

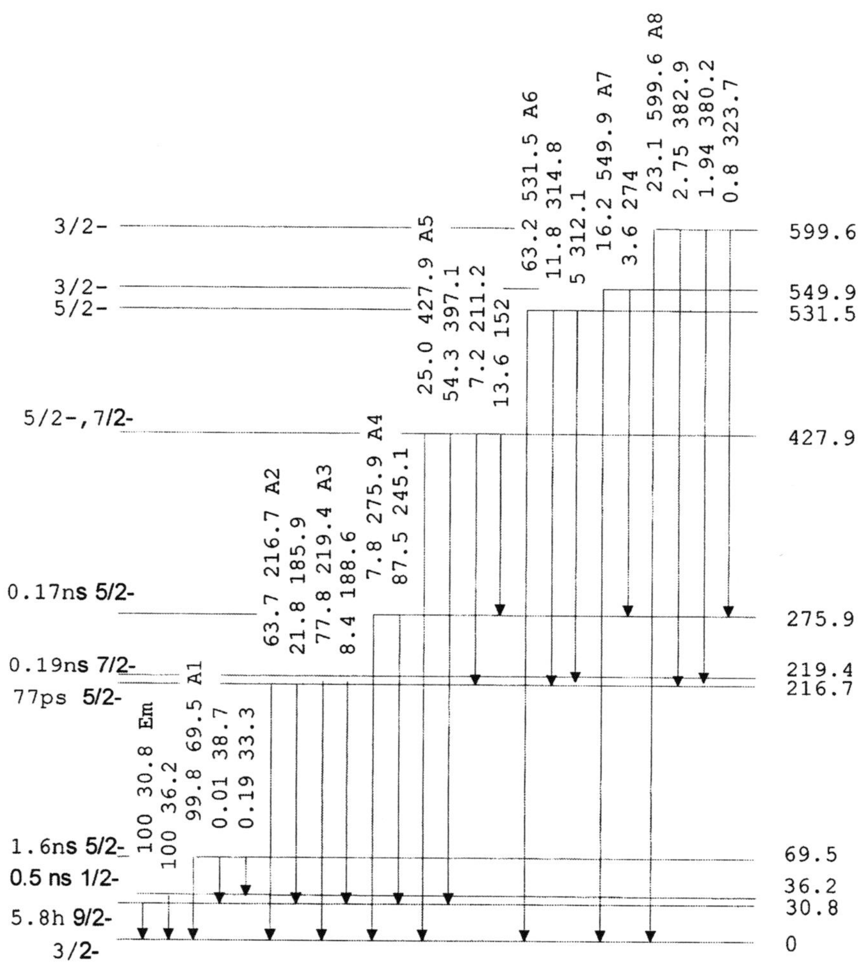

Fig. 69. Partial decay scheme of ^{189}Os up to 600 keV

Besides nuclear resonance absorption (NRA), the first level has long been assumed to also be excited by the NEET process, via virtual photon exchange due to the recombination of vacancies produced in the *K*-shell by incident X-rays. However, contribution from competing NRA cannot be separated, and the claimed dominance or even existence of the NEET process has remained unconfirmed. Also Lakosi et al. [95La3], [95La4] studied the photoexcitation of low-lying nuclear levels of ^{189}Os by ^{137}Cs and ^{60}Co sources.

Veres et al. [73Ve3] investigated five even-even nuclei having isomeric states – among these was also ^{190}Os by ^{137}Cs and ^{60}Co sources. Using the 1.38 and 2.75 MeV γ-rays from ^{24}Na (92.5 TBq) it was supposed that, owing to the continuous energy distribution caused by their Compton-scattering in the source material, they would excite any nuclear level below 2.75 MeV which has a finite branching ratio to the isomeric state. No measurable isomer activity was obtained for $^{190}Os^m$. Therefore the only useful information furnished by this experiment was that the upper limit of the activation cross-section per primary γ-quantum could be estimated as 2×10^{-4} microbarn for $^{190}Os^m$. Kaminishi and Kojima [63Ka], using a 6 MeV linear accelerator, did not observe isomer activity of $^{190}Os^m$ in their experiment.

Gangrsky et al. [96Ga] measured the excitation function of ^{190}Os in the 5–15 MeV energy range. They determined a value of 13.2 nb MeV for the integrated cross-section for saturation. Furthermore, no evidence for the observation of the even-even nucleus ^{192}Os by photoexcitation is mentioned in the literature. It is thought that the spin differences are too high between the ground and isomeric states; therefore, given that the de-excitation of the activation levels to the ground state is the most favourable, the effective photon intensities were not enough to produce a detectable quantity of this isomer. The other reason may be that the *bremsstrahlung* component, which is higher than the isomeric level (1.89 MeV) and which is effective for producing the isomer, is comparatively small.

6.24. $^{191,193}Ir$

There are two natural isotopes of iridium (as indicated above) with ground state spins of $3/2^+$ and rates of abundance of 37.3% and 73.7%, respectively. Both of these have isomeric states with spin $11/2^+$, and $E_{m,191}$ = 171.28 keV and $E_{m,193}$ = 80.22 keV, respectively. The half-lives of the ^{191}Ir, and ^{193}Ir are 4.94 seconds and 10.53 days, respectively.

Kaminishi and Kojima [63Ka] irradiated a natural iridium target using the 5 MeV beam energy of a linear accelerator (average beam current: 20–100 μA) and the induced activation rate was 2.1×10^7 counts per minute per gram. In their experiment they established that $^{191}Ir^m$ has a half-life of 4.90±0.05 seconds.

Veres [63Ve2] also irradiated an iridium target using γ-rays from a 48.5 TBq (1310 Ci) activity ^{60}Co source. He estimated the activation energy level and width level from the experimental results for ^{191}Ir. The overall cross-section was 10 nanobarns and the partial width $\Gamma_{\gamma m}$ was 20 μeV. Abrams and Pelekis [66Ab], using $^{116}In^m$ (E_{max} = 2.112 MeV) radiation sources, produced a neutron reactor loop for the activation of the isomers of iridium. They also determined the overall cross-section of 65±15 nb for $^{191}Ir^m$ but measured no isomer activity for $^{193}Ir^m$. Zaparov et al. [73Za] reported integral cross-section values of 55.1 mb eV if the activation level lies in the energy range of 1.17–1.33 MeV, and 24.7

mb eV if the activation level lies in the energy range of 1.00–1.17 MeV for the $^{191}Ir^{m}$. The resonance γ-flux values in these energy ranges were 3.5×10^{6} and 7.8×10^{6} γ-quanta cm^{-1} s^{-1} eV^{-1}, respectively. ^{60}Co (7.8 PBq ~210 kCi) γ-sources were used for the irradiation of a natural iridium target. Veres et al. [73Ve2] also irradiated a natural iridium target with the γ-rays of ^{24}Na radiation sources (92.5 PBq). The overall cross-sections of $^{191}Ir^{m}$ and $^{193}Ir^{m}$ are 340 nb, and 830 nb, respectively.

In the nineties these isomers were reinvestigated [91Ca], [92Co], [93La1], [93Ma1], [95Kh], and [95La1,2,3,4].

Many properties of nuclear energy levels were studied through photoexcitation. *Table 30* shows the main experimental data on the basis of the above-mentioned works.

Table 30. The excitation cross-sections and other data of the ^{191,193} Ir nuclei

Irradiation facility and/or intensity	Excitation E_{exc} (MeV)	Isomers	σ_{int} (mb eV)	Σ_{int} (mb keV)	$g\,\Gamma_0\,\Gamma_m/\Gamma$ (meV)	$\sigma_{experiment}$ (nb)	Refs
4–6 MeV, LINAC	5	^{191}Ir				2.1×10^{7} *	[63Ka]
^{60}Co, 48.5 TBq	1.1	^{191}Ir			0.1	56	[63Ve2]
$^{116}In^{m}$ reactor loop	<2.112	^{191}Ir				65±15	[66Ab]
		^{193}Ir				n.d.$^{a)}$	
^{60}Co, 3.5×10^{5} $^{b)}$	1.17–1.33	^{191}Ir	55.1		0.022		[73Za]
7.8×10^{6} $^{b)}$	1.00–1.17		24.7		0.007		
^{24}Na, 92.5 PBq	<2.75	^{191}Ir				340	[73Ve2]
		^{193}Ir				830	
Four accelerators,	4	^{191}Ir		76±6.1			[91Ca2]
0.5–11 MeV	6	^{191}Ir		360±9.6			
6 MeV LINAC	1.2±0.3	^{191}Ir		1.8±0.5			[92Co]
	2.5±0.1	^{191}Ir		25±3			
	3.0±0.15	^{191}Ir		50±5			
	4.3±0.2	^{191}Ir		300±40			
Microtron M30	8	^{191}Ir		9.7±1.1 $^{c)}$			[93Ma1]
	10	^{191}Ir		15.4±2.2 $^{c)}$			
Calculated	7–9	^{191}Ir		17 $^{c)}$			
		^{193}Ir		15 $^{c)}$			[93La1]
^{137}Cs, 67.4 TBq	0.659	^{191}Ir	66.2±64			20±19	[95La1]
	0.598	^{193}Ir	5.15±2.5			4.9±2.3	[95La4]
^{60}Co, 2.78 PBq	<1.33	^{193}Ir				107±9	[95La4]
4 MeV LINAC	4, end-	^{191}Ir		400		57000 $^{d)}$	[95La2]
	point	^{193}Ir		172±26		25.4±3.3 μb	[96La]
^{137}Cs, 67.4 TBq	0.659	^{191}Ir	7.5±7.3			20	[95Kh]

* 2.1×10^{7} cpm/g; a) not detected; b) resonance flux in γ/cm^{2} s eV; c) in mb MeV; d) for flux >1 MeV; e) see [95La1], too

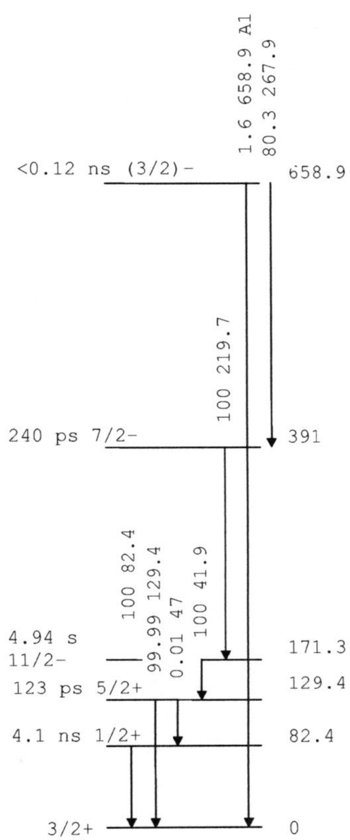

Fig. 70. Partial excitation and γ-decay scheme of ^{191}Ir. A single activation level can be identified up to 662 keV

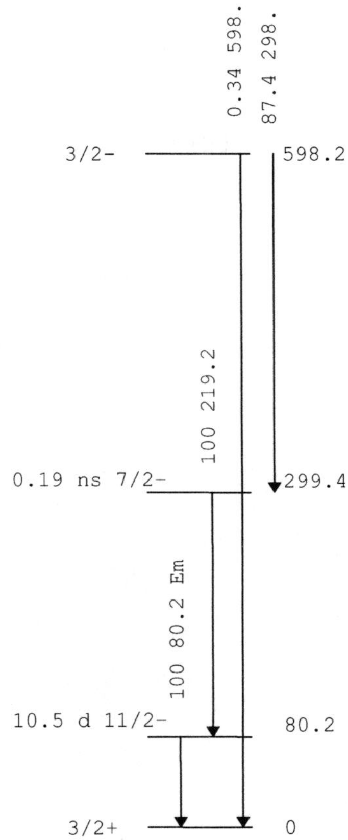

Fig. 71. Partial excitation and decay scheme of ^{193}Ir. A single activation level can be identified up to 662 keV

The investigations concentrated on the activation levels which usually lie in the energy range from 1.1 to 2.75 MeV (corresponding to the principal γ energies of the exciting radioactive sources), and in the higher energy range of 2 to 10 MeV of *bremsstrahlung*.

On the basis of the above-mentioned data and the eighth edition of the Table of Isotopes [96Fi] the most likely photo-excitation and γ-decay schemes were constructed for the ^{191}Ir and ^{193}Ir nuclei (see *Fig. 70–71*). Based on an experiment with a ^{137}Cs source (662 keV), direct transition to the ground state and with cascade transitions to the isomer having been observed, it was easy to identify a single activation level at 658.9 keV, as indicated in *Fig. 70.* There should be more activation levels up to 1.33 MeV (primary energy of ^{60}Co), even not identified. It can be seen from the Table of Isotopes [96Fi] that from 659 keV to 1 MeV there are excited levels having transitions to the ground state –

e.g., 747.8 keV–(5/2)$^+$, 762.55 keV–3/2$^+$, and 935.4 keV–(7/2)$^+$. However, from these levels there is no transition to the isomeric state.

The situations are very similar in ^{193}Ir, too. The increasing values of the overall cross-section with excited energy $E_{exc} < 0.662$ MeV (^{137}Cs), $\sigma_{eff} = 20$ nb; and if $E_{exc} < 1.33$ MeV (^{60}Co), $\sigma_{eff} = 107$ nb; and $E_{ex} < 2.75$ MeV (^{24}Na), $\sigma_{eff} = 340$ nb, then this suggests that there are additional activation levels in the higher energy ranges.

However, some details have emerged which indicate that it would be worth studying even the lower-lying states that are involved in γ-laser research.

6.25. ^{195}Pt

There are six natural isotopes of platinum and only ^{195}Pt (with a spin of 1/2$^-$, and a natural abundance of 33.8%) has an isomeric state. The spin, level energy, and half-life of ^{195}Pt^m are 13/2$^+$, 259.3 keV, and 4.02 days, respectively.

As mentioned above, Kaminishi and Kojima [63Ka] carried out fundamental experiments for the application of the photoexcitation process for activation analysis. The authors also irradiated a natural platinum target with 5 MeV beam energy of a linear accelerator (average beam current: 20–100 μA), and they observed very weakly induced isomer activity (3000 cpm/g) of ^{195}Pt^m.

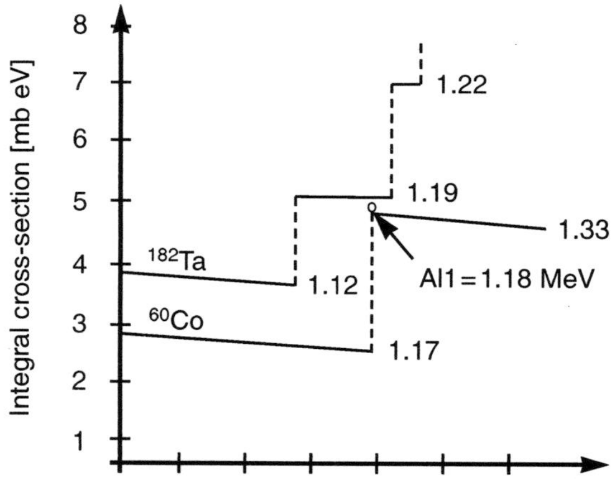

Fig. 72. Integral cross-section values of ^{195}Pt, corresponding to different activation levels as determined by ^{60}Co and ^{182}Ta γ-rays, following the scattered spectrum calibration. The energy of the first activation level is 1.18±0.01 MeV

Veres [63Ve2] irradiated a platinum target using γ-rays from a 48.5 TBq (1310 Ci) activity ^{60}Co source and he measured the 64 keV Pt_{KX}, the 99 keV– and the 130 keV-lines from the γ-decay spectrum of the isomer. It was supposed that the threshold energy of the photoexcitation was about 1.1 MeV and then the number of γ'-quanta scattered into 1 eV at the threshold energy was found to be 1.7×10^{-7} eV^{-1}. So, from the measured (2 nb) overall cross-section the estimated integral cross-section was found to be 12 mb eV and the partial width level of 1.1 MeV was about 4×10^{-6} eV. Abrams and Pelekis [66Ab], using a $^{116}In^{m}$ γ-source, produced a neutron reactor loop for the photoexcitation of $^{195}Pt^{m}$; they reported a value of 2 nb for the cross-section. Zaparov et al. [73Za] determined the integral cross-section values for $^{195}Pt^{m}$. It was found to be σ_{int} = 2.38 mb eV if the activation level lies in the energy range of 1.17–1.33 MeV, and σ_{int} = 1.07 mb eV if the activation level lies in the energy range of 1.00–1.17 MeV. The resonance γ-flux values in these energy ranges were 3.5×10^{6} and 7.8×10^{6} γ-quanta cm^{-1} s^{-1} eV^{-1}, respectively. A ^{60}Co (7.8 PBq ~210 kCi) γ-source was used for the irradiation of the natural platinum target. In an experiment [73Ve2] the authors also irradiated a natural platinum target with γ-rays of ^{24}Na source (92.5 PBq) and measured an overall cross-section value of 328 nb for $^{195}Pt^{m}$.

Veres and Pavlicsek [78Ve2], [80Ve1] carried out experiments for the determination of the activation level energy of ^{195}Pt using the calibration of the scattered resonance γ-flux of two (^{60}Co, and ^{182}Ta) radiation sources. The spectral flux density of the radiation sources used was determined by calculation, and checked by measurement of the isomer activation. Following this spectral-flux density calibration, it was possible to determine an integral cross-section value set on the basis of the measured isomer activity for different energies in those cases where the activation level is unknown (^{195}Pt) but where the isomer activity was induced by both radiation sources. These cases are indicated in *Fig. 72.* Changes in the cross-section values with increasing energy can be seen for 1.17 MeV at ^{60}Co and for 1.19 MeV at ^{182}Ta. The integral cross-section value at 1.12 and 1.19 MeV for ^{182}Ta is in agreement within the error limit for the value of ^{60}Co 1.17 MeV. Thus it is clear that the activation level lies between 1.17 and 1.19 MeV. As a consequence, an activation level of E_{A1} = 1.18±10 keV was rendered probable for ^{195}Pt. Smith et al. [77Sm] excited many levels of ^{195}Pt by neutron pickup reaction; among these a level of 1189±6 keV was found and this proved to be in good agreement with the activation level determined by γ-activation. From the value of the integral cross-section (5.4 mb eV) the lifetime of this level was also calculated and it was found to be 2.7±1 ps.

If the curves for the two irradiators *(Fig. 72)* and the measured equilibrium decay rates are used, two functions (σ_{int} and the activation level) can be determined whose intersection or contact points give the values of integral cross-

Table 31. Some excitation cross-sections data for ^{195}Pt

E_{exc} (MeV)	σ_{eff} (nb)	σ_{int} (mb eV)	Σ_{int} (mb keV)	Refs
1.1	2	12		[63Ve2]
< 2.112	2			[66Ab]
1.17–1.33		2.38		[73Za]
1.00–1.17		1.07		
< 2.75	328			[73Ve]
1.18±0.01		5.4		[80Ve]
4			30±1.6	[91Ca2]
6			140±22	
Calc. 9–11			4400	[93La1]
4 endpoint	8.29±1.1 μb		55.8±8.8	[96La]

section and the energy of the activation level for the unknown activation level. Note, however, that the intersection or meeting of these two curves requires that the ratio of the two spectral functions be a function of E_γ.

The method is therefore viable only when either the primary lines of the source in the irradiators are quite different in energy (so that the activation level energy has a quite different location in the Compton-scattered spectrum of the two sources), or when one or both of the sources contain a number of strong lines in the vicinity of the activation level.

For calibration of the above-mentioned method ^{60}Co and ^{182}Ta radiation sources were used for the comparison of the integral cross-sections. The calibration was performed at the 1.078 MeV activation level of ^{115}In. For the given irradiation arrangements the integral cross-section value for ^{115}In^{m}, determined both by ^{60}Co and ^{182}Ta, was in agreement within the error limit. Calculating the resonant Compton-scattered spectral distribution for the ^{60}Co and ^{182}Ta sources, it was possible to determine an integral cross-section set on the basis of the measured isomer activity.

From the data of the Table of Isotopes [96Fi] it can be seen that the possibility of the appearance of the activation level below 1 MeV is very small, because the spin differences are very large between the ground state and the excitation level. Therefore the cascade transitions via – most probably – *E*1, *E*2, *M*2, and *M*2 transitions to the isomeric states are very unlikely.

Carroll et al. [91Ca2] used four accelerators as sources of *bremsstrahlung* to expose a sample of platinum, and endpoint energies covered the range from 0.5 to 11 MeV. No evidence was found for non-resonant processes of excitation. The resonant photoexcitation of the isomer is 30±1.6 mb keV at 4 MeV endpoint energy and it is 140±22 mb keV at 6 MeV endpoint energy. Lakosi and Sáfár

[93La1] estimated a saturated integral cross-section of 4.4 mb MeV above neutron separation energy. Lakosi et al. [96La] excited the isomer of $^{195}Pt^{m}$ using *bremsstrahlung* of a 4 MeV, 25 μA linear accelerator, and the authors reported 8.29±1.1 μb for a flux >1 MeV; this was for the effective cross-section and for the integrated cross-section of 55.8±8.8 mb keV at 4 MeV endpoint energy. The main data of the experiments are shown in *Table 31.*

6.26. ^{197}Au

The ^{197}Au isotope is the only stable isotope of gold. The spin and parity of the ground state is $3/2^+$. The energy, spin and half-life of the isomeric state are 419.15 keV, $11/2^-$, and 7.73 seconds, respectively.

In 1944 Wiedenbeck observed strong activity when gold was irradiated with high energy X-rays. This activity was attributed to a metastable state of ^{197}Au produced by a (γ, γ') reaction [45Wi]. The period evaluated from a decay curve extending over one minute, was found to be 7.5±0.5 s. The energy level system of gold has been studied in the region from the threshold to 3.3 MeV. From the excitation curve of the isomeric activity the energy of the isomeric state was found to be approximately 250 keV and higher activation levels which

Table 32. Photoexcitation data for ^{197}Au

E_{exc} (MeV)	σ_{eff} (nb)	σ_{int} (mb eV)	Σ_{int} (mb keV)	Γ_{pprt} (μb eV)	Ref.
1.1	0.7	4.1		1	[63Ve2]
< 1.57	87.5				[73Ve2]
< 2.75	9800				
< 2.112	2.6±0.6				[66Ab]
< 2.75	10000±200				[69Ab]
1.22		5		2	[73Za]
0.658 β^+	99±17 μb				[83Vi]
Reviewed	200±17 μb				[88Vi]
4			26±0.4		
6			130±0.64		[91Ca2]
1.7±0.3			0.7±0.3		[92Co]
2.5±0.1			5±0.5		
3.2±0.15			45±5		
4.2±0.2			200±40		
8			7600±1000		
10			11900±2100		[93Ma1]
7–9 Calc.			15000		[93La1]
4	58000		440		[95La2]

combined with the isomer state are found at 1.22±0.03, 1.68, 2.15, 2.56 and 2.97 MeV, respectively.

Veres [63Ve2], using 48.5 TBq ^{60}Co radiation sources, measured the half-life of the isomeric state as 7.2±1 seconds, and he determined the overall and integrated cross-section values of 0.7 nb and 4.1 mb eV at 1.1 MeV energy, respectively; the partial level width of $\Gamma_{\gamma m}$ (of the 1.1 MeV excited energy level) was 1×10^{-6} eV. In the [73Ve2] experiments the authors irradiated a gold target with ^{142}Pr (E_{max} = 1.57 MeV) and ^{24}Na (E_{max} = 2.75 MeV) radiation sources and the overall cross-section values were found to be 87.5 nb and 9.8 μb, respectively.

Abrams and Lakosi [69Ab] irradiated a gold target with $^{116}In^{m}$ (E_{max} = 2.112 MeV) and ^{24}Na (E_{max} = 2.75 MeV) radiation sources and they reported 2.6±0.6 nb and 10±2 μb for the overall cross-section, respectively.

Zaparov et al. [73Za] excited the isomer of ^{197}Au with a ^{60}Co radiation source and they reported an integral cross-section value of 5 mb eV, and a partial level width of 2×10^{-6} eV for the activation level. According to the authors, the most probable activation level is at 1.22 MeV, with $7/2^{+}$ or $5/2^{-}$ spin and parity.

Vishnevsky et al. [83Vi], using ^{64}Cu positron sources (from 0.26 TBq to 0.9 TBq) for irradiating a gold target, reported a partial decay scheme in which they supposed two (1.1 and 1.22 MeV) activation levels. The overall cross-section was found to be 99±17 μb. But for the resonant and non-resonant parts of the cross-sections they reported $\sigma_{res.}\sim10^{-22}$ cm^2 and $\sigma_{nonres}\sim10^{-25}$ cm^2, respectively. These values are far too large and there is no agreement either over the experiment or with regard to the theoretical results. The authors [88Vi] reviewed the experimental and theoretical results of the excitation of the low-lying states of eight nuclei, among them the ^{197}Au. This was done by the annihilation of the positrons and atomic electrons and it was considered to be due to the non-radiating annihilation of positrons. In this case $\sigma_{eff(1)}$ = 99±17 μb and $\sigma_{eff(2)}$ =200±40 μb. They estimated the activation level energy of 1.2 MeV; this moves to the ground state to a degree of 99% and only to a degree of 1% to the isomeric state. The main experimental data are cumulated in *Table 32.*

Discrete-level information for ^{197}Au consists of only 22 levels above the isomeric state, ranging from 0.409 MeV to 1.5 MeV [96Fi]. Unfortunately, the two possible levels (1.12 and 1.22 MeV), from which the nucleus is able to move both to the ground state and to the isomeric state, do not have spin and branching ratios. Therefore *Fig. 73* gives only a small amount of information.

In the 1990s many researchers investigated the production of the isomer by *bremsstrahlung* of linear accelerators in the 0.5–15 MeV energy range.

Carroll et al. [91Ca2], using four accelerators to expose the gold target and endpoint energies, covered the range from 0.5 to 11 MeV. No evidence was

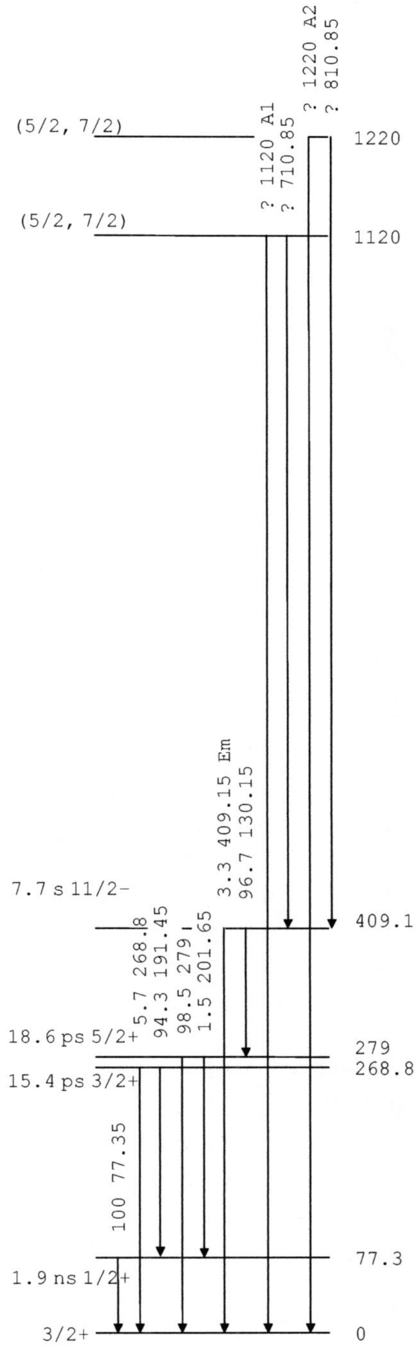

Fig. 73. Possible decay scheme of ^{197}Au

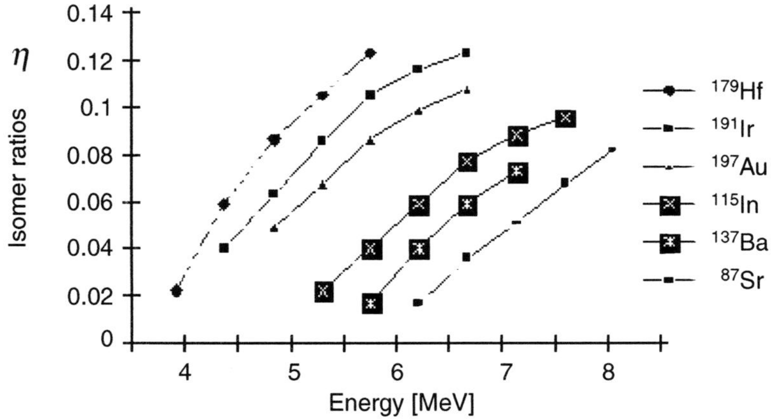

Fig. 74. Energy dependence of the isomer ratios [93Ma1]

found for a non-resonant process of excitation. For irradiation of 4 and 6 MeV, the integrated cross-sections are 26±0.4 mb keV and 130±0.64 mb keV, respectively.

Collins et al. [92Co] measured the integrated cross-section values of $^{197}Au^{m}$ at four different energies (1.7±0.3; 2.5±0.1; 3.2±0.15 and 4.2±0.2 MeV) and these were found to be 0.7±0.3 mb keV, 5±0.5 mb keV, 45±5 mb keV and 200±40 mb keV, respectively.

Mazur et al. [93Ma1] measured the cross-section of γ inelastic scattering with excitation of the isomer state of ^{197}Au at 4–15 MeV energies in ΔE = 0.5 MeV steps. They reported integral cross-sections of 7.6±0.1 mb MeV at 8 MeV excitation energy and of 11.9±2.1 mb MeV at 10 MeV excitation energy. The authors studied the energy dependence of the isomer ratios ($\eta = \sigma_m/\sigma_{tot}$) below the ($\gamma$, n) threshold. Increase of isomer ratios, with an increase of mass number in the 77–197 range, is in qualitative agreement with the predictions of the statistical model *(Fig. 74)*.

Lakosi and Sáfár [93La1] estimated the saturated integral cross-section of 15 mb MeV. Lakosi et al. [95La2], using a 4 MeV, 25 μA linear accelerator, measured the integral cross-section 440 μb MeV for 4 MeV endpoint energy. The effective cross-section was found to be 58 μb for a flux >1 MeV.

6.27. ^{199}Hg

There are seven natural isotopes of the element mercury and only the ^{199}Hg nucleus (with spin 1/2–, and a natural abundance rate of 16.87%) has an isomeric state. The spin, level energy, and half-life of ^{199}Hg^{m} are $13/2^{+}$, 532.48 keV, and 42.6 min, respectively.

Wiedenbeck in 1945 [45Wi] produced the activity of ^{199}Hg^{m} by bombarding mercury with X-rays of a maximum energy of 2.6 MeV. The decay curve indicates a 43-minute period. The relatively small activity prohibited the obtaining of a complete excitation curve at that time and this indicates that the isotope possessing the metastable state was of a relatively low natural abundance.

Veres [63Ve2], using a 48.5 TBq ^{60}Co radiation source, determined the overall cross-section as 0.05 nb and the integrated cross-section as 0.3 mb eV at 1.1 MeV energy. The partial level width, $\Gamma_{\gamma m}$, was estimated to be 1×10^{-6} eV at a 1.1 MeV excited energy level. Veres at al [73Ve2] irradiated mercury with radiation sources of ^{24}Na of 92.5 PBq (2.5 kCi) and the overall cross-section was determined to be 42 nb below the excitation energy of 2.75 MeV.

Law and Iddings [69La] studied the photoactivation of 13 elements, among them mercury. They irradiated 110 grams of HgO powder (17.1 g ^{199}Hg) with 185 PBq (5 kCi), and 1.11 PBq (30 kCi) ^{60}Co radiation sources, but they recorded no isomer activity for ^{199}Hg^{m}.

Table 33. A few photoexcitation data of ^{199}Hg

E_{exc} (MeV)	σ_{eff} (nb)	σ_{int} (mb eV)	Σ_{int} (μb MeV)	Γ_{part} (10^{-6} eV)	Ref.
1.1	0.05	0.3		0.1	[63Ve2]
<2.75	42				[73Ve2]
<2.112	0.08				[66Ab]
<1.49	<0.015		$<7\times10^{-4}$	<0. 4	[67Bo]
1.49	$0.41\pm^{0.2}_{0.15}$		$(6.9\pm^{3}_{2})\times10^{-2}$	41	
1.61	$0.35\pm^{0.15}_{0.1}$		$(9\pm^{4}_{3})\times10^{-2}$	63	
1±0.02			$(1.1\pm^{2}_{0.4})\times10^{-5}$		[69Bo]
1.34±0.01			$(5.1\pm^{10}_{2})\times10^{-4}$		
1.38±0.01			$(2.2\pm^{3.8}_{0.7})\times10^{-3}$		
1.42±0.01			$(7.9\pm^{13.6}_{2.8})\times10^{-3}$		
1.53±0.02			$0.12\pm^{0.21}_{0.03}$		
1.7±0.02			$0.2\pm^{0.49}_{0.12}$		
1.17–1.33		0.25			[73Za]
4			2600±100		
6			14000±420		[91Ca2]
8			550±100		
10			660±150		[93Ma1]
7–9			9500		[93La1]

Zaparov et al. [73Za], using a 7.77 PBq (210 kCi) ^{60}Co radiation source for the irradiation of a high quantity (1360 g) liquid mercury target, determined the resonance flux of 7.8×10^6 γ-quanta per cm^2 s eV in the energy range of 1.17–1.33 MeV at the place of the target and during irradiation. The integral cross-section was found to be 0.25±0.1 mb eV.

Abrams and Pelekis [66Ab] irradiated mercury with γ-rays (E_{max} = 2.112 MeV) of $^{116}In^m$ radiation sources produced in a neutron reactor loop and reported 0.08 nb for the overall cross-section.

Booth and Brownson [67Bo] excited ^{199}Hg by photons and electrons at energies below 3 MeV, and the threshold of the excitation was found to be at 1.41 MeV. Below the threshold the integral cross-section was found to be $< 7 \times 10^{-4}$ μb MeV and it was 6.9×10^{-2} μb MeV at 1.49 MeV (1/2, 3/2, 5/2) and 9×10^{-2} μb MeV at 1.61 (1/2, 3/2, 5/2) MeV excitation energy.

Boivin et al. [69Bo] determined integral cross-sections and partial level widths in the 1.0–1.7 MeV excitation energy range.

In the 1990s, in several laboratories, the photoexcitation of ^{199}Hg was studied in a higher energy range (3–15 MeV). As mentioned above, [91Ca2], [93La1], [93Ma1] studied numerous isomers of the stable elements, among them the ^{199}Hg nucleus. The main values, with other experimental data, are shown in *Table 33.*

6.28. ^{204}Pb and ^{207}Pb

There are four natural isotopes of lead, but only ^{204}Pb (with spin 0^+, and a natural abundance rate of 1.4%), and ^{207}Pb (with spin $1/2^-$, and a natural abundance rate of 22.1%) have isomeric states. The spins, level energies, and half-lives of the isomeric states of the two nuclei ($^{204}Pb^m$ and ^{207}Pb) are 9^-, 2189.79 keV, 67.2 minutes, and $1/2^-$, 1633.368 keV, 0.805 seconds, respectively.

Kaminishi and Kojima [63Ka] irradiated targets of even-even nuclides such as ^{90}Zr, $^{178,180}Hf$, ^{190}Os, and ^{204}Pb with 5 MeV *bremsstrahlung* of a linear accelerator produced by a 4–6 MeV electron beam; however, they observed no isomeric activity in their experiment. The $^{204}Pb^m$ isotope emits hard γ-rays that are easily detectable, but in spite of this the authors did not observe any isomer activity. Law and Iddings [69La1] studied the photoactivation of 13 isomers, among them $^{204}Pb^m$. They observed no isomeric activity either. They irradiated 20 grams of a lead target with 185 PBq (5 kCi), and 1.11 PBq (30 kCi) ^{60}Co radiation sources, but they observed no isomer activity in the case of $^{204}Pb^m$. It has already been mentioned that Veres et al. [73Ve3] investigated the possibilities of the photoexcitation of five even-even (^{90}Zr, ^{178}Hf, ^{180}Hf, ^{190}Os and ^{204}Pb) nuclei. Using the 1.38 and 2.75 MeV γ-rays from ^{24}Na (92.5 TBq) it was supposed that, owing to the continuous energy distribution caused by their Compton-scattering in the source material, they would excite any nuclear level below 2.75 MeV, this having a finite branching ratio to the isomeric state. The extremely low probability of the photoexcitation process may in a large part be ascribed to the structures of these nuclides and the interaction with photons.

Gangrsky et al. [96Ga] detected a value of 218 mb MeV for the saturated cross-section of ^{204}Pb.

A photoexcitation experiment of ^{207}Pb^{m} was carried out by Abrams and Pelekis [68Ab] with ^{116}In^{m} radiation sources (E_{max} = 2.112 MeV) produced in a neutron reactor loop; no isomer activity was observed. Lakosi and Sáfár [93La1] calculated the integral cross-section of ^{207}Pb^{m} and they reported a value of 9.5 mb MeV.

6.29. ^{210}Bi^{m}

^{209}Bi is the last stable nucleus to be produced in the slow process of nucleosynthesis by a (n, γ) reaction. When it captures a neutron, both the ground state of ^{210}Bi ($t_{1/2}$ = 5 d) and its (long-life) isomer ($t_{1/2} = 3 \times 10^6$ y) are formed.

In the process of developing a preliminary cross-section set Gardner and Gardner [88Ga1] determined some of the parameters, such as those for the level densities, from the analyses of discrete-level information. They found that this information must be as complete as possible. It should not only consist of those levels obtained through experimentation, but must also be supplemented with by model calculations, and with known theoretical level structures.

The authors [88Ga1] also considered the response of ^{210}Bi states to incident photons. The 9^- isomeric state is not produced by a photon incident in the 1^- ground state. Furthermore, as shown in *Fig. 75,* neither is it destroyed to any great extent by photons. Photons on the 9^- (m_1) target state mainly produce that state itself. The next state which is most probably excited is the 7^- state (433.5 keV, 58 ns); this normally decays completely back to the 9^- state. The destruction reactions – i.e., those that lead to the ground state or the 5^- state (439.2 38 ns) – have reactions lower by several orders of magnitude than the earlier reactions. Thus the 9^- state should not be populated by photons.

One question remains: what role, if any, is played by the photo-production peaks below 1 MeV? These primarily reproduce the 9^- state, but the peaks around 0.3, 0.65, and 0.9 MeV also produce the 7^- state to degrees ranging between 15% to 30% of the time. It is conceivable that the photon flux is sometimes great enough to allow a two-step process to occur, where the 7^- state is produced and then absorbs before decaying to the 9^- state.

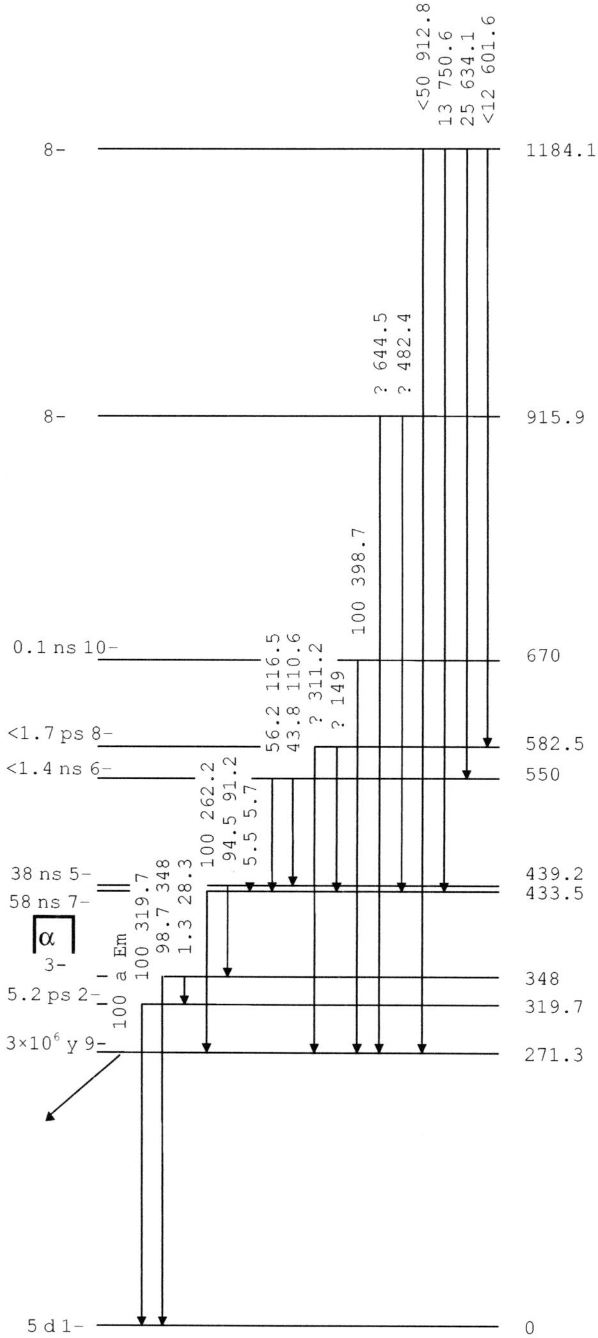

Fig. 75. A partial and decay scheme of ^{210}Bi

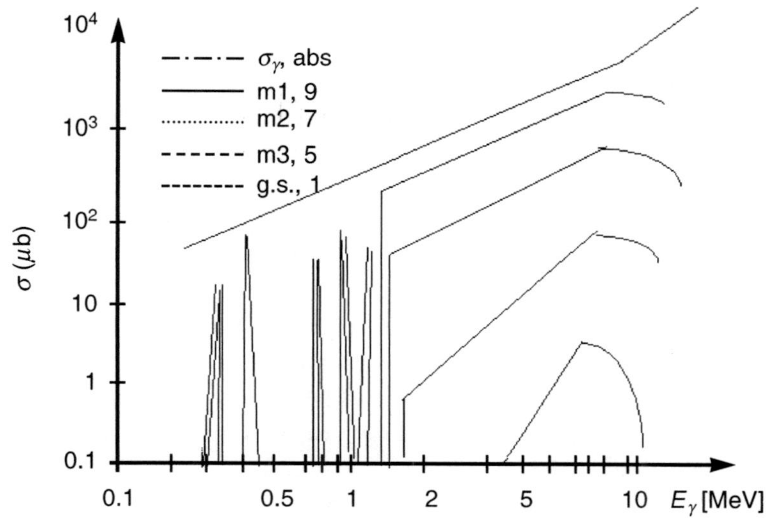

Fig. 76. The photo-production of various states of ^{210}Bi by photons incident on the 9^- isomer. The $\sigma_{g,\,abs}$ curve is the calculated total photon absorption cross-section

For photons above 1.1 MeV, the 7^- state produces similar amounts of the 9^- and 5^- states, but very little of the 1^- ground state is produced directly. Similar calculations were made with the 5^- level (i.e., as with the target), and the photo-excitation results were not greatly different from those for the 7^- state *(Fig. 76)*.

It would seem, therefore, that the photo-destruction of the 9^- isomer of ^{210}Bi is not a serious problem, even in high photon fluxes where multiple photon absorption may occur.

6.30. ^{235}U

There are three natural isotopes of the element uranium. Two of them, the nucleus ^{234}U (with ground state spin 0^+, and a natural abundance of 0.055%), and the nucleus ^{238}U (with ground state spin 0^+, and a natural abundance of 99.2745%) have no isomeric state. The existence of a very low-energy isomeric state in ^{235}U was discovered independently by Huizenga et al. [57Hu], and by Asaro and Perlman [57As]. The 26±0.16 min isomeric decay is thus expected to go – by an E3 transition – from a presumed $1/2^+$ isomeric level to the $7/2^-$ ground state of ^{235}U. The transition energy has the exceptionally low value of 76.8 eV, and proceeds by internal conversion in the outermost electron sub-shells: most probably the $(6p_{1/2})^2$, $(6p_{3/2})^4$ and $(6d)$ shells of uranium. These sub-shells, and consequently the isomeric decay rate, can be perturbed by changes in the valence electron sub-shells $(5f)^3$, $(6d)$, $(7s)^2$. The conversion electron (CE) spectrum of this transition has been studied by several authors [57Fr], [66Ma], [68Mc], [70Ne], and it has been shown that, indeed, the decay rate changes by several per cent with the metal into which $^{235}U^{m}$ is implanted [74Ne].

Due to the very low energy of the $^{235}U^{m}$ conversion electrons, special detection systems are required for these measurements; furthermore, appropriate techniques must be used for the source preparation, so as to reach adequate counting rates and reproducible decay rate and energy distribution results. The effects to be detected are small, thus all secondary phenomena have to be carefully taken into account.

Morita [73Mo] produced a theoretical study for a new mechanism for nuclear excitation by considering the de-excitation of electronic states of the atom. It was found that the probability of this process is extremely small in most of the cases, but it will be appreciable if the interaction energy is approximately equal to the energy mismatch $E_N - (E_1 - E_2)$, where E_N is the nuclear excitation energy. Although this is a very rare event, it is not impossible to detect it since the decay of this state creates the 26-minute isomeric. This is subsequently de-excited by the internal conversion process.

Although there are no (γ, γ') experiments to be found in the literature, it can be seen from the partial decay scheme of *Fig. 77* that seven activation levels can be found below 0.5 MeV excitation energy.

There are numerous activation levels above the 500 keV energy range as well. Therefore it would be worth carrying out photoexcitation experiments with X- or γ-rays.

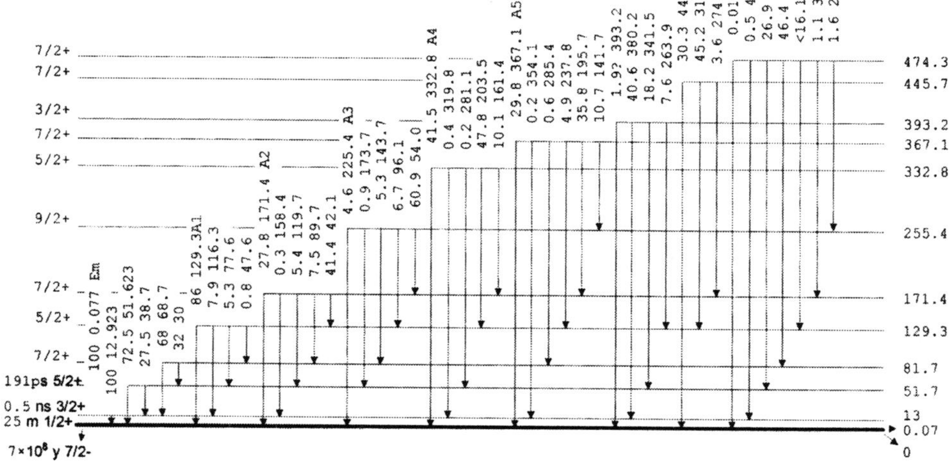

Fig. 77. A partial excitation and decay scheme of ^{235}U, up to 500 keV excitation energy

7. Conclusions

In the evaluation of the photoexcitation method, in addition to data compilations, it has mainly been the applicability of methods studied in the fields of nucleosynthesis and gamma laser developments which have been examined here.

It is difficult to predict the extension of these methods for understanding the play of some isomers in the process of nucleosynthesis. One of the continuing questions in nuclear astrophysics concerns the origin of the naturally occurring $^{180}Ta^{m}$ isomer. It has been indicated that a likely astrophysical site for the photonuclear synthesis of nuclei exists in white dwarf-accreting surface layers of hydrogen-rich material, leading to a nova outburst. Unique astrophysical information is encoded in gamma-ray lines: not only do their energies indicate the presence of specific nuclei, electron-positron pairs and burst line properties such as intensities and profiles, but they also contain information on natural rates of abundance, bulk velocities, gravitational potentials, densities, temperatures and so on. The recent results from gamma-ray observatories have demonstrated that γ-line spectroscopy is a fascinating research field which has the potential to open entirely new windows onto astronomy. There is hope that these cumulated experimental and calculated results of photoexcitation cross-sections and other data will be some help in this respect.

Research on the development of a gamma-ray laser has followed a cyclical pattern in the past 40 years, with a marked abundance of concepts and approaches. The quest for a gamma laser resembles that for controlled nuclear fusion in many respects. Both have been pursued for many years, each with an effort proportional to their generally perceived respective pay-offs. Despite many negative conclusions, researchers have found far more positive proposals and there are more potential approaches to graser programs today than ever before. The fundamental precept is that the very high densities of energy storage should be found within the nucleon. However, in rare cases selection rules sufficiently inhibit the coupling of the particle motion to the electromagnetic field for the energies stored for tens and even thousands of years in those special nuclei. Such isomers are natural sources of high energy densities. It can be seen that the energy storage of the best of isomers, $^{178}Hf^{m2}$ (half-life: 31 years) is more than a gigajoule per gram. Since one of the key transitions for the output of electromagnetic radiation has a life-time 70 ns after triggering, the power density available from that material is (1.3×10^9 J $g^{-1}/7 \times 10^{-8}$ s^{-1}) ~ 1.86×10^{16} J g^{-1} s^{-1}, or about 18.6 peta-watt per gram. The triggered release of stored energy into an electromagnetic wave works and works well. Production of this type of ultra-high energy density material is still in its infancy, although remarkable progress has already been made. For example, more than 10^{16} nuclei of $^{178}Hf^{m2}$ are already available, but these are not yet mass-separated and carrier-free. This is an amount from which it should be possible to obtain a target of more than 10^{15} nuclei with an almost 100% inversion (i.e. for future experiments).

The proof that isomeric populations can be dumped by pumping them with intense pulses of X-rays has been followed by the demonstration that the strengths for doing so are fragmented into relatively few discrete transitions of great intensity. These are precisely the necessary conditions for the concentration of useful pump power into the selective excitation of nuclear populations, while dissipating power degraded to heat in much larger volumes. There are no *a priori* obstacles to the realisation of a gamma laser. A gamma laser seems feasible if the right combination of energy levels occurs with some real material.

References[1]

[21Ha]. O. Hahn, *Über eine neue radioaktive Substanz im Uran,* Chem., Berichte, **54**, 1131

[29Hu]. E. P. Hubble, *A Relation Between Distance and Radial Velocity Among Extra-Galactic Nebulae,* Proc. Nat. Acad. Sci. **15**, 168

[35Ku]. V. Kurchatov, I. Kurchatov, L. Moskovsky, L. Rusinov, *Sur un cas de radioactivité artificielle provoquée par un bombardment de neutrons, sans capture du neutron,* Compt. Rend. **200**, 1201

[36Br]. G. Breit, E. Wigner, *Capture of Slow Neutrons,* Phys. Rev. **49**, 519

[36We]. C. F. Weizsäcker, *Metastabile Zustände der Atomkerne,* Naturwiss., **24**, 813

[37Pe]. C. Perrier, E. Segré, *Radioactive Isotopes of Element 43,* Nature,140,193

[38Go]. M. Goldhaber, R. D. Hill, L. Szilard, *Nuclear Physics,* Nature, **142**, 521

[38Ol]. O. Oldenberg, *Artificial Radioactivity of Tantalum,* Phys. Rev. **53**, 35

[39Co]. G. B. Collins, B. Waldman, E. M. Stubblefield, M. Goldhaber, *Nuclear Excitation of Indium by X-Rays,* Phys. Rev. **55**, 507

[39Go]. M. Goldhaber, R. D. Hill, L. Szilard, *Radioactivity Induced by Nuclear Excitation,* Phys. Rev. **55**, 47

[39La]. W. E. Lamb, *Capture of Neutrons by Atoms in Crystal,* Phys. Rev. **55**, 190

[39Po]. B. Pontecorvo, A. Lazard, C.R. Hebd. *Isomérie nucléaire produite par les rayons X du spectra continou,* Seances Acad. Sci. **208**, 99

[41Co]. G. B. Collins, B. Waldman, *Energy Distribution of Continuous X-Rays from Nuclear Excitation,* Proc. of the APS in Phys. Rev. **59**, 109

[41Gu]. E. Guth, *Radiative Transition Probabilities in Heavy Nuclei; Excitation of Nuclei by X-Rays,* Phys. Rev. **59**, 325

[41Fe]. J. Feldmaier, G. B. Collins, *Excitation of Nuclei by X-Rays,* Phys. Rev. 59, 937

[43Ko]. W. Korsunsky, G. Ivanov, J. Phys. **7**, 129

[43Wa]. B. Waldman, M. L. Wiedenbeck, *The Nuclear Excitation of Indium by X-Rays and Electrons,* Phys. Rev. Lett. **63**, 60

[45Gu]. E. Guth, *Nuclear Spectroscopy and Inelastic Scattering of Particles by Nuclei,* Phys. Rev. Lett. **68,** 279; *Nuclear Spectroscopy and Energy Distribution of Charged Particles Inelastically Scattered by Nuclei,* Phys. Rev. Lett. **68**, 280

[45Wi]. M. L. Wiedenbeck, *Nuclear Isomerism of Gold,* Phys. Rev. Lett. **67**, 53; *The Nuclear Excitation of Silver and Cadmium,* Phys. Rev. **67**, 92; *The Nuclear Excitation of Rhodium,* Phys. Rev. Lett. **67,** 267; *The Excitation of Heavy Nuclei,* Phys. Rev. **68**, 1; *The Nuclear Excitation of Krypton and Rhodium,* Phys. Rev. **68**, 237

[1] In the references [21Ha] the two first numbers denote the year of the publication; so, for example, the first reference given above – [21Ha] – indicates that the work was published in 1921.

[47Du]. I. V. Dunworth, B. Pontecorvo, *Excitation of Indium 113 by X-Rays,* Proc. Camb. Phil. Soc., **43**, 123
[48Al]. R. A. Alpher, H. Bethe, G. Gamow, *The Origin of Chemical Elements,* Phys. Rev. Lett. **73**, 803
[49Mi]. W. C. Miller, B. Waldman, *An Investigation of Bremsstrahlung by Means of the Nuclear Isomerism of Indium,* Phys. Rev. **75**, 425
[51Br]. H. N. Brown, W. L. Bendel, F. J. Shore, R. A. Becker, *Disintegration of Ta^{180},* Phys. Rev. **84**, 292
[51Sc]. E. Schillinger, W. C. Miller, B. Waldman, *An Investigation of Bremsstrahlung by Means of the Nuclear Isomerism, of Indium II,* Phys. Rev. **83**, 320
[52Pr]. R. D. Present, S. C. Chen, *Nuclear Disintegration by Positron-K Electron Annihilation,* Phys. Rev. **85**, 447
[53Go]. J. Goldemberg, L. Katz, *High Energy Gamma-Gamma Cross Section of In^{115},* Phys. Rev. **90**, 308
[53Si]. C. S. del Rio y Sierra, V. L. Telegdi, Phys. Rev. **90**, 439
[54Ha]. G. Harbottle, *Activation of Nuclear Isomers by Gamma Rays,* Nucleonics, **12**, 65
[54Mi]. W. Miller, J. W. Motz, C. Cialella, *Thick Target Bremsstrahlung Spectra for 1.00-, 1.25-, and 1.400-MeV Electrons,* Phys. Rev. **96**, 1344
[55Bu]. J. L. Burkhardt, E. J. Winhold, T. H. Dupree, *Photoexcitation of the Isomeric State of Indium-115,* Phys. Rev. **100**, 199
[55El]. J. P. Elliot, T. H. R. Skryme, *Centre-of-mass Effects in the Nuclear Shell Model,* Proc. Roy. Soc. **A232**, 561
[55Wh]. F. A. White, T. L. Collins, F. A. Rourke, *New Naturally Occurring Isotope of Tantalum,* Phys. Rev. **97**, 566
[56Bo]. O. V. Bogdankevich, L. E. Lazareva, F. A. Nikolaev, *Inelastic Scattering of Photons on the Nucleus of ^{115}In,* Zh. Eksp. Teor. Fiz. **31**, 405
[56Ik]. N. Ikeda, K. Yoshihara, *Radioactivation by the (γ,γ)-Reaction I-II (in Japanese),* Radioisotopes, **5, (1),** 11; **5,(2)**, 11
[56Me]. L. Meyer-Schützmeister, V. L. Telegdy, *Inelastic Nuclear Scattering of Photons by Au^{197},* Phys. Rev. **104**, 185
[56Ro]. H. R. Robl, *Kern-Anregung durch Electronen,* Nucl. Phys. **2**, 641
[57As]. F. Asaro, I. Perlman, *Isomeric State of Uranium-235,* Phys. Rev. Letter to the Editor, 318
[57Bi]. G. R. Bishop, R. Wilson, in *Handbuch der Physik,* J. Springer-Verlag, Berlin, Vol. **40/I**, 281
[57Bu]. E. M. Burbidge, G. R. Burbidge, W. A. Fowler, F. Hoyle, *Synthesis of the Elements in Stars,* Rev. Mod. Phys., **29**, 547
[57Fr]. M. S. Freedman, F. T. Porter, F. Wagner, P. P. Day, *Transition in 26-min U^{235m} of Less Than 23 Electron Volts,* Phys. Rev. **108**, 836
[57Ha]. C. L. Hammer, M. G. Stewart, *Isomeric Transition in the Rare Earth Elements,* Phys. Rev. **106**, 1001
[57Hu]. J. R. Huizenga, D. W. Engelkemeir, *27-Minute Isomer of U^{235},* Phys. Rev. **107**, 319
[57Mi]. J. W. Mihelich, B. Harmatz, *Some New Isomeric Transition in Rare Earth Nuclei,* Phys. Rev. **106**, 1232
[58Ik]. N. Ikeda, K. Yoshihara, Radioisotopes, *Radioactivation by the (γ, γ)-Reaction III–VII,* Radioisotopes, **7**, 11; **7**, 17; **7**, 92; **7**, 196; **8**, 17 (in Japanese)

[59Ca]. A. G. W. Cameron, *Photobeta Reactions in Stellar Interiors,* Astrophys. J. **130**, 452
[59Me]. F. R. Metzger, *Resonance Fluorescence in Nuclei,* Prog. Nucl. Phys. **7**, 54
[59Ik]. N. Ikeda, K. Yoshihara, *Radioactivation by the (γ,γ)-Reaction VIII–IX,* Radioisotopes, **8**, 24, **8**, 238, (in Japanese)
[60Bo]. E. C. Booth, *Nuclear Resonance Scattering of Bremsstrahlung,* Nucl. Phys. **19**, 426
[60Wi]. D. Wilkinson, *Analysis of Gamma Decay Data,* in Nuclear Spectroscopy ed. by Fay Ajzenberg-Selove, Academic Press, New York and London, 852
[60Yo]. K. Yoshihara, *Studies on the Radioactivation by the Cobalt-60 γ-Ray Sources I–III,* Isot. Radiat. **3**, 276; **3**, 464; **3**, 472, (in Japanese)
[61Lu]. H. R. Lukens, J. W. Otvos, C. B. Wagner, *Formation of Metastable Isomers by Photo-activation with the Van de Graaff Accelerator,* Int. J. Appl. Radiat. Isot. **11**, 30
[61Ot]. J. W. Otvos, V. P. Guinn, H.R. Lukens, C. D. Wagner, *Photoactivation and Photo-neutron Activation Analysis,* Nucl. Instr. Meth. **11**, 187
[61Ri]. L. A. Rivlin, USSR Patent #621265 (Appl. Jan. 10, 1961)
[61Yo]. K. Yoshihara, *Studies on the Radioactivation by the Cobalt-60 γ-Ray Sources IV–V,* Isot. Radiat. **4**, 93; **4**, 102 (in Japanese)
[62Ba]. G. C. Baldwin et al., General Electric Co. Reports, No. 62GL22 and No. 62GL176
[62Bo]. E. C. Booth, K. A. Wright, *Nuclear Resonance Scattering of Bremsstrahlung,* Nucl. Phys. **35**, 472
[62Hu]. J. R. Huizenga, R. Vandenbosch, *Photoexcitation of the Isomeric States of* Cd^{111} *and* In^{115} *with Mono-energetic Gamma Rays,* Nucl. Phys. **34**, 457
[62Ka]. Y. Kagan, V. A. Maslow, *Mössbauer Effects in Mono-, and Di-Atomic Cubic Grids,* Zs. Eksp. Teor. Fiz. **41**, 1296
[62Ne]. P. E. Nemirovsky, Yu. V. Adamchuk, *Nuclear Resonance Scattering of Bremsstrahlung,* Nucl. Phys. **58**, 551
[63Ba]. G. C. Baldwin, J. P. Neissel, J. H. Terhune, L. Tonks, *On the Possibility of Maser Action Between Nuclear States,* Tr. Am. Nucl. Soc.**6**, 178
[63Bi]. G. E. Bizina, A. G. Beda, N. A. Burgov, A. V. Davydov, *Resonance Excitation of an Isomer State in* Ag^{107} *with a Mean Lifetime of 63 sec,* Zs. Eksp. Teor. Fiz., **45**, 1508
[63Ch]. B. V. Chirikov, *Kinetics of Induced Mössbauer Radiation,* Zh. Eksp. Teor. Fiz. **44**, 2016
[63Ka]. T. Kaminishi, C. Kojima, *Production of Nuclear Isomers from Stable Nuclides with Bremsstrahlung,* Jap. J. Appl. Phys. **2**, 399
[63Ma]. D. Marcuse, *Maser Action Without Population Inversion,* Proc. IEEE **51**, 849
[63Ri]. L. A. Rivlin *(Bragg reflections in perfect crystals),* Vopr. Radioelektron, **6**, 42*)*; and *(Derivation of the conditions of a start-oscillation),* Vopr. Radioelektron, **6**, 60 (in Russian)
[63Va]. V. Vali, W. Vali, *Induced γ-Ray Emission,* Proc. IEEE, **51**,182; **51**, 1228
[63Ve1]. Á. Veres, *Photoactivation of Cadmium-111m and Indium-115m by Cobalt-60 Irradiation,* Int. Appl. Radiat. Isot., **14**, 123
[63Ve2]. Á. Veres, *Photoactivation of Nuclear Isomers by* Co^{60} *Irradiation,* Acta Phys. Hung., **16**, 261 (in Russian)
[64Cl]. D. D. Clayton, *Cosmoradiogenic Chronologies of Nucleosynthesis,* Ap. J. **139**, 637
[64En]. C. Engelmann, Report CEA-R 2559, Saclay, France
[64Ka]. Yu. Kagan (Theoretical Study on the *Temperature Broadening of the Mössbauer-Lines*), Zs. Eksp. Teor. Fiz., **47**, 366 (in Russian)
[65Ch]. B. T. Chertok, E. C. Booth, *Nuclear Excitation by 1 MeV to 3 MeV Electrons,* Nucl. Phys. **66**, 230

[65Pe]. A. A. Penzias, R. W. Wilson, *A Measurement of Excess Antenna Temperature at 4080 Mc/s,* Astroph. J. **142**, 419
[66Ab]. I. A. Abrams, L.L Pelekis, I. Ya. Taure, *(Excitation of Nuclear Isomers by γ-Rays of ^{116m}In Source),* Latv. PSK, Zinat, **66(1)**, 3; **66(6)**, 3; **66(6)**, 95; **68(1)**, 19 (in Russian)
[66Ca]. Y. Cauchois, Y. Heno, M. Boivin, *Photoactivation Nucléaire de l'argent naturel, de ses isotopes 107 et 109 et du cadmium naturel,* Phys. Nucl., C. R. Akad. Sc. **262**, B-503
[66Ma]. H. Mazaki, S. Shimizu, *The Transition Energy of $^{235}U^{m}$,* Phys. Lett., **23**, 167
[66Pö]. W. P. Pönitz, *A Gamma-Ray Cascade Model for the Calculation of Average γ-Ray Multiplicities and Isomeric Cross Section Ratios,* Z. Phys.**197**, 262
[66Ve1]. Á. Veres, M. Ozsgyáni, *Detection of a Cd^{113} Isomer with the Help of Gamma Radiation from Co^{60}* (Yad. Fiz.), Sov. J. of Nuclear Physics, **Vol. 3, Nr. 5**, 587,
[66Ve2]. Á. Veres, *Application of Nuclear Photoeffect in Activation Analysis,* Proc Anal. Chem. Conf. Budapest, 360
[66Ve3]. Á. Veres, I. Pavlicsek, *Activation of the Nuclear Isomer Indium-113^{m} by Gamma Gays of Cobalt-60,* JARI, **17**, 69
[66Ve4]. Á. Veres *(Studies and Measurement of Photon Spectra of ^{60}Co γ-Rays),* Magyar Fizikai Folyóirat, **14**, 143 (in Hungarian)
[67Ab]. I. A. Abrams, L. L. Pelekis, *Gamma-Activation Dosimeter,* Latv. PSK, Zinat, **67(2)**, 3; *Gamma-Activation Method for Determining of Br, Ag and In in Some Materials,* Latv. PSK, Zinat, **67(5)**, 45 (in Russian)
[67Bo]. E. C. Booth, J. Brownson, *Electron and Photon Excitation of Nuclear Isomers,* Nucl. Phys. **A98**, 529
[67Cl]. D. D. Clayton, M. E. Rassbach, *Termination of the s-Process,* Ap. J. **148**, 69
[67Ku]. Kulenkampff in *Handbook of Physics,* ed. E. U Condom, McGraw-Hill, New York
[67Li]. R. E. Lingenfelter, R. Ramaty, in *High Energy Nuclear Reactions in Astrophysics*, ed. B. S. P. Shen, Benjamin Inc. P. 99
[67St]. M. M. Stautberg, J. J. Kgaushaar, B. W. Ridley, *Nuclear Structure Studies of Sr^{88} and Y^{89} Using 19 MeV Proton Scattering and the $Sr^{88}(He^{3},d)$ Reaction,* Phys. Rev. **157**, 977
[67Ve]. Á. Veres, I. Pavlicsek, M. Ozsgyáni, *Measurement of High-Activity Cobalt-60 Solid Sources by The Application of Nuclear Photoeffect,* Proc. Standardisation of Radionuclides, IAEA, 453
[67Zo]. W. Zobel et al., O. N. R. L. 3506 and 4183 unpublished Reports
[68Al]. W. J. Alston III, H. H. Wilson, E. C. Booth, *Resonance Fluorescence and Absorption Measurements of Transition Strengths in Odd-Mass Nuclei,* Nucl. Phys. **A116**, 281
[68Cl]. D. D. Clayton, *Principles of Stellar Evolution and Nucleosynthesis*, 1968, McGraw-Hill, Book Company, New York, Chapter 7, 518
[68Mc]. E. C. McCullough, *A Detailed Study of those Factors Affecting the Useful Lifetime of the Continuous Channel Electron Multiplier,* Nucl. Inst. Meth. **60**, 351
[68Pe]. G. A. Peterson, J. Alster, *Nuclear Structure of ^{88}Sr and Y^{89} by Inelastic Electron Scattering,* Phys. Rev. 166, 136
[69Ab]. I. A. Abrams, L. Lakosi, *Na^{24} Gamma-Ray Excitation of Nuclear Isomers (in Russian),* Latv. PSK, Zinat, **6**, 3
[69Bo]. M. Boivin, Y. Cauchois, Y. Heno, *Photoactivation Nucléaire du ^{77}Se, $^{107,109}Ag$, ^{11}Cd, ^{115}In at ^{199}Hg,* Nucl. Phys., **A137**, 520
[69La]. J. J. Law, F. A. Iddings, *A Study of Photoactivation with a ^{60}Co Source,* J. Radioanal. Chem. **3**, 53

[69Ra]. S. Ramchandran, J. A. McIntyre, *Spins and Widths of Energy Levels in the 5–6 MeV Region,* Phys. Rev. **179**, 1152
[69Ve]. Á. Veres, I. Pavlicsek, *Nuclear Photoactivation Analysis by Means of an 80-kCi ^{60}Co γ-Radiation Source,* J. Radioanal. Chem. **3**, 25
[69Vet]. J. I. Vette, J. L. Matteson, D. Gruber, Proc. Rome Conference, on X-ray astronomy
[69Wa]. G. J. Wasserburg, D. N. Schramm, J. C. Huneke, *Nuclear Chronologies for the Galaxy,* Ap. J. **157**, L91
[70Au]. J. Audouze, *Some Nucleosynthetic Effects of Energetic Proton Fluxes on Stellar Surfaces and on the Solar System,* Astron. & Astrophys. **8**, 436
[70Fo]. W. A. Fowler, H. Reeves, F. Hoyle, *Galactic Cosmic Ray Origin of Li, Be and B in Stars,* Nature **226**, 727
[70La]. L. Lakosi, Á. Veres, *Measurement of a Multicurie Activity ^{24}Na Source by Isomeric Activation,* Nucl Instr. Meth. **83**, 213
[70Mo]. R. Moreh, S. Slomo, A. Wolf, *Radiative Widths, Spins, and Parities of Nuclear Levels in the 5–9 MeV Region Excited by Neutron Capture γ Rays,* Phys. Rev. **C2**, 1144
[70Ne]. M. Neve de Mevergniens, *Energy of the Isomeric Transition in ^{235}U,* Phys. Lett.**32B,** Nr 6, 482
[70Ry]. C. Ryter, H. Reeves, E. Gradsztajan, J. Audouze, *The Energetic of L-Nuclei Formation in Stellar Atmospheres and its Relevance to X-Ray Astronomy,* Astron. Ap. **8**, 389
[70Ve]. A. Veres, I. Pavlicsek, *Measurement of the Activation Cross Section of ^{176}Lu γ-Rays from a ^{60}Co Radiation Source,* Acta Phys. Hung. **28**, 419
[71Al]. L. Allen, G. I. Peters, *Amplified Spontaneous Emission: I. The Threshold Condition,* J. Phys. **4 A**, 238; *II. The Connection with Laser Theory,* J. Phys. **4 A**, 377; *III. Intensity and Saturation,* J. Phys. **4 A**, 564
[71Bo]. M. Boivin, Y. Cauchois, Y. Heno, *Photoactivation Nucléaire de ^{115}In,* Nucl. Phys. **A176**, 626
[71Gu]. S. C. Gujrathi, J. M. D'Auria, *The Decay of ^{180}Lu to the Levels of ^{180}Hf,* Nucl. Phys. **A161**, 410
[71Lu]. G. J. Lutz, *Photon Activation Analysis – A Review,* Anal. Chem.**43**, 93
[71Me]. F. R. Metzger, *Model-dependent and Model-independent Determinations of the B(E2) for the First 2^+ State in ^{90}Zr,* Bull. Am. Phys. Soc. **16**, 562
[71Wi]. F. C. Williams, *Particle-Hole State Density in the Uniform Spacing Model,* Nucl. Phys. **A166**, 231
[72Ka]. T. Kato, Y. Oka, *Yields of Photonuclear Reactions for Photon-Activation Analysis with High-Energy Bremsstrahlung,* Talanta, Pergamon Press, North. Ireland, **19**, 515
[72Me1]. F. R. Metzger, *The B(E2) for the First 2^+ State in ^{90}Zr,* Nucl. Phys. **A182**, 213
[72Me2]. P. Meyer, *Activation Analysis Methods Applied to Coins,* Review in "Meth. Chem Met. Inv. Anc. Coinage", Royal Num. Soc. Pub. **No. 8**, 183
[72Mu]. T. Mukoyama and S. Shimizu, *Nuclear Excitation by Positron Annihilation,* Phys. Rev. **C5**, 95
[73Ba]. G. C. Baldwin, in Laser Interaction and Related Plasma Phenomena Edited by H. J. Schwarz, H. Hora, Plenum, New York, **3B**, 875; **4A**, 249; **4A**, 259
[73Bo]. C. D. Bowman, *Application of Photonuclear Physics,* Proc. First Int. Conf. Photonuclear Reactions, (Asilomar, USA), No. 730401, 1351
[73Go]. V. I. Gol'dansky, Yu. M. Kagan, Zh. Eksp. Teor. Fiz. **64**, 90 (in Russian)

[73Le]. V. S. Letokhov, *Narrow-Frequency Tuneable γ-Ray Nuclear Resonances Induced by Laser Radiation,* Phys. Rev. Lett. **30**, 729

[73Mo]. M. Morita, *Nuclear Excitation by Electron Transition and Its Application to Uranium-235 Separation,* Prog. Theor. Phys. **49,** 1574

[73Ot]. K. Otozai, R. Arakawa, M. Morita, *Experimental Evidence for Nuclear Excitation by Electron Transition in ^{189}Os,* Prog. Theor. Phys. **50**, 1771

[73Te]. E. Teller, Proc. First Int. Conf. Photonucl. React. Appl. (Asimolar, USA), No. 730401, 1357

[73Ve1]. Á. Veres, *Rapid Determination of Erbium Using Gamma Rays from ^{24}Na Radiation Sources,* J. Radioanal. Chem. **16**, 605

[73Ve2]. Á. Veres, I. Pavlicsek, M. Csûrös, L. Lakosi, *Nuclear Excitation by γ Rays of Radioactive Sources,* Acta Phys. Hung. **34**, 97

[73Ve3]. Á. Veres, I. Pavlicsek, M. Csûrös, *Possibilities of Photo-Excitation of Even-Even Nuclear Isomers,* Proc. First Int. Conf. Photonuclear Reactions (Asilomar, USA), No. 730401, 307

[73Vey]. A. Veyssiere, H. Beil, R. Bergere, P. Carlos, A. Lepretree, *A Study of the Photofission and Photoneutron Processes in the Giant Dipole Resonance of ^{232}Th, ^{238}U, and ^{237}Np,* Nucl. Phys. **A19**, 45

[73Za]. E. A. Zaparov, Yu. N. Koblik, B. C. Mazitov, G. A. Tadyuk, *Direct Reactions and Isomer Transitions*, FAN Uzb. SSR Tashkent, 5 (in Russian)

[74Ar]. B. Arad, G. Ben-David, *Resonance Fluorescence of Excited Nuclear Levels in the Energy Range 5-11 MeV,* Annual Review of Nucl. Sci. **24**, 35

[74By]. J. Byrne, G. I. Peters, L. Allen, *Stimulated Emission from Nuclei,* Appl. Opt. **13**, 2499

[74Di]. V. F. Dmitriev, S. V. Shuryak, *On the Feasibility of the γ-Laser,* Zh. Eksp. Teor. Fiz. **67**, 494

[74Ed]. T. W. Edwards, T. G. Harrison, *Photoneutron Mechanism for the Production of Technetium-99 in the Interior of Evolved Stars,* Ap. J. **187**,313

[74Ha]. T. G. Harrison, T. W. Edwards, *Low Temperature Photoneutron Sources for Stellar Nucleosynthesis,* Ap. J. **187**, 303

[74La]. L. Lakosi, M. Csûrös, Á. Veres, *Isomer Activation Data from Experiments by High Activity ^{60}Co Sources,* Nucl. Instrum. Methods, **114**, 13

[74Me]. F. R. Metzger, *Resonant Scattering of Bremsstrahlung by ^{90}Zr,* Phys. Rev. **C9**, 1525

[74Ne]. M. Neve de Mevergniens, P. Del Marmol, *Effect of the Oxidation State on the Half-life of $^{235}U^m$,* Phys. Lett. **49B**, No. 5, 428

[75Au]. J. Audouze, J. W. Truran, *p-Process Nucleosynthesis in Postshock Supernova Envelope Environments,* Astrophys. J. **202**, 204

[75Ha]. T. G. Harrison, *A Photonuclear Mechanism for Promethium Synthesis Including Technetium,* Ap. Lett. **16**, 87

[76Oc]. A. P. Ochkur, *Gamma ray methods in ore geology* (ed. Nedra, Leningrad, 1976, in Russian)

[76Wa]. R. A. Ward, M. J. Newman, D. D. Clayton, *s-Process Studies: Branching and the Time Scale,* Ap. J. Suppl. **31**, 33

[77Ch]. A. Chattopadhyay, *Optimal Use of Instrumental Neutron and Photon Activation Analyses for Multielement Determinations in Sewage Sludges,* J. Radioanal. Chem., **37**, 785

[77Ri]. L. A. Rivlin, *Nuclear Radiative Transitions in the Laser Radiation Field,* Kvant. Elektron (Moscow), **4**, 676; *On Stimulation of Two-Quantum Annihilation of Relativistic Electrons and Protons,* Kvant. Elektron (Moscow), **5**, 2497 (in Russian)

[77Sm]. G. R. Smith, N. J. Giacomo, M. L. Munger, R. J. Peterson, *Neutron Pickup from ^{194}Pt, ^{196}Pt and ^{198}Pt,* Nucl. Phys. **A290**, 72

[77Ve]. Á. Veres, *Determination of Selenium Content of Animal Food-Stuff Using Photon Activation Analysis,* J. Radioanal. Chem. **38**, 155

[77Wa]. R. A. Ward, *The Importance of Long-Lived Isomeric States in s-Process Branching,* Ap. J. **216**, 540

[77Yo]. K. Yoshihara, A. Fujita, T. Shiokawa, *Appearance Energies of Indium Complex Compounds Determined by the $^{115}In(\gamma, \gamma')^{115m}In$ Reaction,* J. Inorg. Nucl. Chem., **39**, 173

[78Ge]. Germán E., Veres Á., *(A method for non-destructive assay of spent fuel assemblies by isomer activation of ^{115}In, Hungarian Patent)*, Szabadalmi Közlöny **83**, 548, T/15369

[78Ha]. T. G. Harrison, *The Low-Temperature Photonuclear Nucleosynthesis of the Bypassed (p-) Nuclei in Degenerate Hydrogen Burning Zones and Its Relationship to Nova Outbursts,* J. Ap. Suppl. **36**, 199

[78Hs]. S. T. Hsue, *Methods for the Non-destructive Assay of Irradiated Nuclear Fuels for Safeguards,* Atomic Energy Review, **16**, 89

[78Le]. C. M. Lederer, V. S. Shirley, Table of Isotopes, 7th Ed., New York, 1978

[78Pa]. D. C. Palmer et al., *Low-lying Levels of ^{129}Xe and ^{131}Xe,* J. Phys. G. Nucl. Phys. 1143

[78Ot]. K. Otozai, R. Arakawa, T. Saito, *Nuclear Excitation by Electron Transition in ^{189}Os,* Nucl. Phys. **A297**, 97

[78Ve1]. Á. Veres, K. Yoshihara, N. Ikeda, *Excitation of Nuclear Isomers Applied to Activation Analysis, Measurement of High Activity Source and Hot Atom Chemistry,* Radioisotopes, **27**, 43

[78Ve2]. Á. Veres, I. Pavlicsek (*Determination of Activation-Level Energy of ^{195}Pt by the Calibration of Spectral Flux Density),* Prikladnaya Yadernaya Spektroskopiya, **Ed. 8**, 283 (in Russian)

[79Ar]. B. Arad, S. Eliezer, Y. Paiss, *Nuclear "Anti-Stokes" Transition by Laser Radiation,* Phys. Lett. **A 74**, 395

[79Ba1]. G. C. Baldwin, *Time-Domain Spectroscopy of Recoilless Gamma Rays,* Nucl. Instr. Meth. **159**, 309

[79Ba2]. G. C. Baldwin, J. C. Solem, *Maximum Density and Capture Rate of Neutrons Moderated from a Pulsed Source,* Nucl. Sci. Eng. **72**, 281; **72**, 290

[79Br]. Ph. Breban et al. *Mode des Possibilities d'Utilisation Analitiqe des Isomers Nucleaires Produits par Reactions (γ, γ') Entre 6–8 MeV* Nucl. Instr. Meth. **158**, 205

[79Ir]. A. D. Irving, P. D. Forsyth, I. Hall, D.G.E. Martin, *The Properties of Low-lying Levels of ^{129}Xe and ^{131}Xe,* J. Phys. G.Nucl.Phys.,1595

[79Iz]. I. Izawa, G. Yamanaka, *Production ^{235}U by Nuclear Excitation by Electron Transition in a Laser Plasma,* Phys. Lett. **88B**, 59

[79Mo]. S. A. Moszkovszki, *Theory of Multipole Radiation*, in K. Sieghbahn, *α-, β-, and γ-Ray Spectroscopy,* Amsterdam, North Holl. Publ. Co., 863–886 pp.

[79Vi]. I. N. Vishnevsky, V. A Zheltonozhsky, B. P. Svyato, V. V. Trishin, *Excitation of Nuclei ^{115}In, ^{113}In, and ^{111}Cd Induced by the Positron Annihilation* Izv. Ak. Nauk, **43**, 2142 (in Russian)

[79Vy]. V. I. Vyshotskii, *On the Possibility of Non-Threshold γ Amplification in a System of Polarized Nuclei,* Zh. Eksp. Teor. Fiz. **77**, 492 (in Russian)

[79Wa1]. E. Warde, R. Seltz, G. J. Costa, D. Magnac, C. Gerardin, *The Ground State of ^{180}Ta Is Not 8^+,* J. Phys. Lett. **40**, L1

[79Wa2]. Y. Watanabe, T. Mukoyama, S. Shimizu, *Nuclear Excitation of ^{115}In by Positron Annihilation with K-Shell Electrons,* Phys. Rev. **C19**, 32

[80Be]. H. Beer, F. Käppeler, *Neutron Capture Cross Sections on ^{138}Ba, $^{140,142}Ce$, $^{175,176}Lu$ and ^{181}Ta at 30 keV Prerequisite for Investigation of the ^{176}Lu Cosmic Clock,* Phys. Rev. **C21**, 534

[80Ry]. T. B. Ryves, *The Decay Scheme of ^{180m}Ta,* J. Phys. G: Nucl. Phys. **6**, 763

[80Sh]. K. S. Sharma, R. J. Ellis, V. P. Derenchuk, R. C. Barber, *Atomic Mass of ^{180}Ta a Naturally Occurring Isomer,* Phys. Lett. **91B**, 211

[80St]. A. S. Shtan, *Development of Nuclear and Physical Methods of Element Analysis of the Substance Composition in the USSR,* J. Radioanal. Chem. **59**, 591 (in Russian)

[80Ve1]. Á. Veres, *Gamma Activation of Nuclear Isomers and Its Applications,* At. Energy Rev. **18**, 271

[80Ve2]. Á. Veres, I. Pavlicsek, *Determination of Certain Trace Element Content by Gamma Activation Analysis,* Proc. Trace Element, Walter de Gruyter & Co., Berlin

[80Wa]. R. A. Ward, W. A. Fowler, *Thermalization of Long-lived Nuclear Isomeric States Under Stellar Conditions,* Ap. J. **238**, 266

[81Al]. J. Allen, G. C. Lowenthal, J. R. de Later, *s-Process Branch at ^{176}Lu,* J. Phys G: Nucl. Phys. **7**, 1271

[81Ba]. G. C. Baldwin, J. G. Solem, V. I. Gol'danskii, *Approaches to the Development of Gamma-Ray Lasers,* Rev. Mod. Phys., **53**, 687

[81Be1]. H. Beer, R. A. Ward, *Neutron-capture Nucleosynthesis of Nature's Rarest Stable Isotopes,* Nature, **291**, 308

[81Be2]. H. Beer, F. Käppeler, K. Wisshak, *^{176}Lu: Cosmic Clock or Stellar Thermometer,* J. Ap. Suppl., **46**, 295

[81Bo]. Yu. N. Bourmistenko, *Gamma-Activation Installation for Fast Determination of Gold and Its Accompanying Elements in Ore Samples,* Isotopenpraxis, **Bd. 17**, 241

[81Co]. C. B. Collins, in: *Proc. Int. Conf. on Lasers '80* (ed. C. B. Collins, STS Press, Maclean, VA, 1981), 524

[81De]. R. A. Dewberry, R. K. Sheline, R. G. Lanier, L. G. Mann, G. I. Struble, *Nuclear Levels in ^{176}Lu,* Phys. Rev. **C24**,1628

[81Lj]. A. Ljubicic, K. Pisk, B. A. Logan, *Photoactivation Investigations with ^{115}In,* Phys. Rev. **C23**, 2238

[81No]. E. B. Norman, *Half-life of ^{180}Ta,* Phys. Rev. **C24**, 2334

[81Ol]. S. Olariu, I. Popescu, C. B. Collins, *Tuning of γ Ray Process with High Power Optical Radiation,* Phys. Rev. **C23**, 50; *Multiphoton Generation of Optical Sidebands to Nuclear Transitions,* Phys. Rev. **C23**, 1007

[81Ra]. R. S. Raghavan, A. P. Mills, *Nuclear Excitation by Positron Annihilation: Comments on Theory vs. Experiment,* Phys. Rev. **C24**, 1814

[81Wa]. Y. Watanabe, T. Mukoyama, R. Katano, *Nuclear Excitation of ^{176}Lu by Positron Annihilation,* Phys. Rev. **C23**, 695

[82Ba]. C. A. Barnes, D. D. Clayton, D. N. Schramm, *Essays in Nuclear Astrophysics,* Cambridge University Press

[82Be1]. H. Beer, *^{176}Lu: Cosmic Clock and Stellar Thermometer,* Astrophys. J. **262**, 739

[82Be2]. H. Beer, R. L. Macklin, *$^{178,179,180}Hf$ and $^{180}Ta(n, \gamma)$ Cross Sections and Their Contribution to Stellar Nucleosynthesis,* Phys. Rev. **C26**, 1404

[82Es]. P. Eskola et al., *Level Structure of ^{83}Kr from the $^{80}Se(\alpha, n\gamma)^{83}Kr$ Reaction,* Phys. Scripta, **25**, 15

[82Kr]. M. Krcmar, A. Ljubicic, K. Pisk, B. A. Logan, M. V. Tar, *Photoactivation of ^{111}Cd,* Phys. Rev. **C25**, 2097

[82Ol]. S. Olariu, I. I. Popescu, C. B. Collins, N. Ceausescu, *Amplification of Gamma Radiation from X-Ray Excited Nuclear States,* Rev. Roum. Phys. **27**, 559

[82Pi]. K. Pisk, M. Krcmar, A. Ljubicic, B. A. Logan, *Nuclear Excitation by the Inelastic Photoelectric Effect,* Phys. Rev. **C25**, 2226

[82Ve]. Á. Veres, I. Pavlicsek, *Determination of Noble Elements in Coins by Gamma Activation Analysis,* J. Radioanal. Chem. **69,** 325

82Wa]. Y. Watanabe, T. Mukoyama, *Nuclear Excitation by Gamma Rays from Cobalt-60,* Nucl. Sci. Eng. **80**, 92

[83Cl]. D. D. Clayton, *Extinct Radioactivities a Three-phase Mixing Model,* Ap. J. **268**, 381

[83De]. R. A. Dewberry, R. A. Naumann, *Two-quasiparticle Excited Neutron Levels in ^{180}Ta,* Phys. Rev. **C28**, 2259

[83Lö]. T. Lönnroth, J. Kumpulainen, C. Tuokko, *One- and Three-Quasiparticle States in $^{127,129,131,133}Xe$ and Their Coexistence With Band Structures,* Phys. Scripta, **27**, 228

[83Vi]. I. N. Vishnevsky et al. *Excitation of Nuclei ^{103}Rh, $^{107,109}Ag$, and ^{197}Au Induced by the Positron Annihilation on the Bound Electron* (Yad., Fiz.), Sov. J. Nucl. Phys. **38**, 2(8), 273

[83Yo]. K. Yokoi, K. Takahashi, *Slow Neutron Capture for $^{180}Ta^m$,* Nature, **305**, 198

[84Co]. C. B. Collins, B. D. Paola, *Tunable Sub-Angstrom Radiation Generated by Anti-Stokes Scattering from Nuclear Levels,* AIP Conf. Proc. No. **119**, 45

[84De]. B. D. DePaola, C. B. Collins, *Tunability of Radiation Generated at Wavelengths Below 1 Å by Anti-Stokes Scattering from Nuclear Levels,* J. Opt. Soc. Am. **B1**, 812

[84Es]. W. Eschner et al. *Measurement of the β-Decay Branch in ^{180}Lu and ^{180m}Hf for Investigation of the r-Process Nucleosynthesis of ^{180m}Ta,* Z. Phys **A317**, 281

[84Ho]. R. Horvat, K. Pisk, B. A. Logan, *Nuclear Excitation During Pair Production,* Phys. Rev. **29C**, 1614

[84Lj]. A. Ljubicic, M. Krcmar, K. Pisk, R. A. Logan, *Mechanism for Nuclear Excitation by Positron Annihilation,* Phys. Rev. **30C**, 2109

[84No]. E. B. Norman et al., *On the Half-Life of ^{180}Ta in Stellar Environments,* Ap. J. **281**, 360

[84Sc]. A. Schenzle et al., *Microscopic Theory of Optical Line Narrowing of a Coherently Driven Solid,* Phys. Rev. **A30**, 325

[85Ak]. J. M. Akkermans, H. Gruppelaar, *Comments on the State Densities and the Transition Rates in the Pre-Equilibrium Excitation Model,* Z. Phys. **A 321**, 605

[85Co]. C. B. Collins, B. D. DePaola, *Observation of Coherent Multiphoton Processes in Nuclear States,* Optics Lett. **10**, 25

[85Cu]. J. B. Cumming, D. A. Alburger, *Search for the Decay of $^{180}Ta^m$,* Phys. Rev. **C31**, 1494

[85El]. R. C. Elton, *X-Ray Lasers,* Handbook of Laser Sci. Techn. CRC Press, Vol. I. 445

[85Ga]. M. A. Gardner, D. G. Gardner, *The Importance of Level Structure in Nuclear Reaction Cross Section Calculations,* ACS Symp. **0097-6156**, 100

[85Ke]. D. Kekez, A. Ljubicic, K. Pisk, B. A. Logan, *Nuclear De-excitation via the Electronic-Bridge Mechanism,* Phys. Rev. Lett., **55**, 1366

[85Kr]. W. F. Krupke, *Types and Comparisons of Laser Sources,* Handbook of Laser Sci. Techn. CRC Press, Vol. I. 3

[85La]. L. Lakosi, Á. Veres, I. Pavlicsek, Zs. Németh, *Gamma and Neutron Activation for ND Assay of Irradiated Fuel Assemblies,* Proc. 7th Symp. ESARDA, Liege, Belgium, 265

[85No1]. E. B. Norman, T. Bertram, S. E. Kellogg, S. Gil, P. Wong, *Equilibration of* $^{176}Lu^{m}$ *During the s-Process,* Astrophys. J., **291**, 834

[85No2]. E. B. Norman, S. E. Kellogg, *Some Effects of High Temperature and Density on Neutron-Capture Nucleosynthesis,* Proc. AIP Symp. **0094-243X**, 753

[85Pi]. K. Pisk, M. Krcmar, A. Ljubicic, A. B. Logan, *Nonradiative versus Radiative Nuclear Excitation in the Positron-K-Electron Annihilation,* Phys. Rev. **326**, 83

[86Ba]. V. G. Batij, E. A. Skakun, Yu. N. Rakivnenko, O. A. Rastrepin, *Determination of the Excitation Cross Section of* ^{180m}Ta *in (p,n) reaction,* Izv. AH, **50**, 913

[86Ke]. S. E. Kellogg, B. Norman, *Search for the Beta Decay of* ^{180}Lu *to* $^{180}Hf^{m}$*,* Phys. Rev. **C34**, 2248

[86Kr1]. M. Krcmar, A. Ljubicic, B. A. Logan, M. Bistrovic, *Photoactivation of Isomeric Levels in* ^{113}In *and* ^{87}Sr*,* Phys. Rev. **C33,** 293

[86Kr2]. M. Krcmar, A. Ljubicic, K. Pisk, B. A. Logan, M. Bistrovic, *Secondary Photon-Atom Contributions to Nonresonant Excitation of* ^{115}In *Isomeric Level,* Fizika, **18**, 171

[86Le]. K. T. Lesko et al., *Isomeric Levels in* ^{180}Lu *and the Nucleosynthesis of* $^{180}Ta^{m}$*,* Phys. Rev. **C34**, 2256

[86Ma]. G. J. Mathews, K. Takahashi, R. Ward, M. Howard, *Stellar Technetium and Niobium Abundances as a Measure of the Lifetime of AGB Stars in the Third Dredge-up Phase,* Astrophys. J. **302**, 410

[86Mü]. I. Müller, *Nucl. Data Sheets,* **49**, 589

[86So]. J. Solem, AIP Conf. Proc. *Inter-level Transfer Mechanism and Their Application to Grasers,* **146**, 22

[86Va]. W. J. Varhue, T. G. Williamson, *Inelastic Photon Scattering by* ^{111}In*,* Appl. Radiat. Isot. **37**, 155

[86Yo]. K. Yoshihara, Zs. Németh, L. Lakosi, I. Pavlicsek, Á. Veres, *Major Mechanism of Photoactivation for the* $^{115}In(\gamma,\gamma'^{115}In^{m}$ *Process by* ^{60}Co *γ-Ray Irradiation,* Phys. Rev. **C33, 7**28

[87An]. J. A. Anderson, C. B. Collins, *Calibration of Pulsed Bremsstrahlung Spectra with Photonuclear Reactions of* ^{77}Se *and* ^{79}Br*,* Rev. Sci. Instrum. **58**, 2157

[87Bi]. I. Bikit et al., *Photoactivation of* $^{111}Cd^{m}$ *without a Nonresonant Contribution,* Phys. Rev. **C35,** 1943

[87Ga]. M. A. Gardner, D. G. Gardner, *The Impact of Calculated Photon-Induced Isomer Production in* ^{176}Lu *on Its Use as a Stellar Chronometer and/or Thermometer,* Proc. Capture Gamma-ray, Leuven, Belgium, 1

[87Kr). M. Krcmar, A. Ljubicic, M. Bistrovic, *Photoactivation Investigations with* ^{113}In *and* ^{87}Sr*,* Nucl. Instr. Meth. **A255**, 99

[87La]. L. Lakosi, I. Pavlicsek, Zs. Németh, Á. Veres, *Photoexcitation of* $^{189}Os^{m}$*,* Proc. Symp. Nucl. Struct. **13**, 568

[87Né]. Zs. Németh, L. Lakosi, I. Pavlicsek, Á. Veres, *Half-lives of* ^{111m}Cd*,* ^{113m}In *and* ^{115m}In*,* Appl. Radiat. Isot, **38**, 63

[87Pi]. K. Pisk, A. Ljubicic, B. A. Logan, *Higher-Order Processes in the Photon-Nucleus Interaction,* Nucl. Instr. Meth. **A255**, 28

[87Ru]. E. Runte et al., *The Decay of the New Isotope* ^{180}Yb *and the Search for the r-Process Path to* ^{180m}Ta*,* Z. Phys. A-Atomic Nuclei, **328**, 119

[87Sh]. A. Shinohara et al., *Nuclear Excitation in ^{189}Os with Synchrotron Radiation,* Nucl. Phys. **A472**, 151

[87Ta]. K. Takahashi, K. Yoko, At. Data and Nucl. Data Tables **36**, 375

[88An]. J. A. Anderson, C. B. Collins, *Calibration of Pulsed X-Ray Spectra,* Rev. Sci. Instrum., **59**, 414

[88Co]. C. B. Collins, C. B. Eberhard, J. W. Glesener, J. A. Anderson, *Depopulation of the Isomeric State $^{180}Ta^m$ by the Reaction $^{180}Ta(\gamma, \gamma')^{180}Ta$,* Phys. Rev. **C37**, 2267

[88Di]. S. S. Dietrich, B. L. Berman, *Atlas of Photoneutron Cross Sections Obtained with Monoenergetic Photons,* At. Data Nucl. Data Tables, **38**, 199

[88Ga1]. D. G. Gardner, M. A. Gardner, *Preliminary Evaluation of the 3×10^6 y Isomeric State of ^{210}Bi in Astrophysical s-Process Nucleosynthesis,* Annual Report, LLNL, US, 53

[88Ga2]. S. Gales, Ch. Stoyanov, A. I. Vdovin, *Damping of High-Lying Single-Particle Modes in Heavy Nuclei,* Phys. Report, Phys. Lett.**166**, 125

[88Ga3]. M. A. Gardner, D. G. Gardner, R. W. Hoff, *The Impact of Calculated Photon-induced Isomer Production in ^{176}Lu on Its Use as a Stellar Chronometer,* J. Phys. G: Nucl. Phys. 14 Suppl. S315

[88Ho]. Yu-Kun Ho, Zheng-Ying Pan, *Laser-induced Isomeric Transitions and Gamma-Ray Laser,* Nucl. Phys. **A 486**, 271

[88Ik]. N. Ikeda, R. Seki, M. Kamemoto, M. Otsuji, Radioisotopes, **37**, 414

[88Kä]. F. Käppeler, *Nuclear Excitation and Stellar Temperature*, J. Phys. G: Nucl. Phys. 14 Suppl. S297

[88La]. L. Lakosi, I. Pavlicsek, Á. Veres, *Photoexcitation of $^{176}Lu^m$ and Related Astrophysical Implications,* Proc. Capt. γ-ray Spectr. S745

[88Né]. Zs. Németh, Á. Veres, *Population of the Metastable State of ^{111}Cd via Photoactivation by ^{60}Co source,* Phys. Rev. **C38**, 533

[88Ve]. Á. Veres, l. Lakosi, I. Almási, Zs. Németh, I. Pavlicsek, *Photoactivation of Nuclear Isomers for Assaying Irradiated Reactor Fuel,* Proc. NDST, Mito, 959

[88Vi]. I. N. Vishnevsky, V. A. Zheltonozhsky, V. M. Kolomiezh *(Excitation of Nuclei by the Annihilation of the Positrons),* Sov. Fiz. Element. Particle and Nuclei **19/2,** 237 (in Russian)

[88Yo]. K. Yoshihara, T. Sekine, Zs. Németh, L Lakosi, Á. Veres, *Formation of $^{99}Tc^m$ by the Nuclear Excitation,* Nucl. Sci. Technology, ed. S. Igarashi, JAERI, Saigon, 1075

[88Zh]. Z. Zhao et al. *New Excited States of the Nucleus ^{129}Xe,* Z. Phys. A- Atomic Nuclei, **331**, 113

[89An]. J. A. Anderson, et al. *Nuclear Photactivation Cross Sections for Short-lived Isomeric States Excited with a 6 MeV LINAC,* Nucl. Instr. Meth. **B40/41**, 452

[89Bi]. Z. M. Bigan, L. E. Lazareva, V. M. Mazur, I. V. Sokolyuk, *Excitation of Isomer States of Nuclei ^{167}Er, ^{179}Hf and ^{183}W in $(\gamma, \gamma')^m$ Reaction,* J. Nucl Phys. (Sov. Yad. Fiz.), **49/4**, 913

[89Ca]. J. J. Carroll et al., *Accelerated Decay of ^{180}Ta and ^{076}Lu in Stellar Interiors Through (γ, γ') Reactions,* J. Ap. **344,** 454

[89Pi]. H. H. Pitz et al. *Systematic Study of Low-lying Dipole Excitations in $^{156,158,160}Gd$ by Photon Scattering,* Nucl. Phys. **A 492**,411

[89Se1]. T. Sekine, K. Yoshihara, Zs. Németh, L. Lakosi, Á. Veres, *A New Detection Method of ^{99}Tc by Nuclear Excitation,* J. Radioanal. Nucl. Chem. **130**, 269

[89Se2]. T. Sekine, K. Yoshihara, I. Pavlicsek, L. Lakosi, Á. Veres, *Measurement of $^{103}Rh^m$ Produced by the $^{103}Rh(\gamma,\gamma')^{103}Rh^m$ Reaction with Liquid Scintillation Counting*, J. Radioanal. Nucl. Lett., **137**, 57

[89Yo]. K. Yoshihara, H. Kaji, T. Sekine, T. Nakajima, T. Mukoyama, *Photonuclear Excitation of ^{103}Rh by Synchrotron Radiation,* IARI, **40**, 491
[90Bl]. J. Blachot, F. Haas, Nuclear Data Sheets **60**
[90Co]. C. B. Collins et al., *Resonant Excitation of the Reaction $^{180}Ta^{m}(\gamma, \gamma')^{180}Ta$,* Phys. Rev. **C42**, R1813
[90Du]. A. P. Dubenskiy, V. P. Dubenskiy, E. A. Boykova, L. Malov, Bull. An, USSR, Ser. Phys. **54**, 166
[90Dz]. L. Z. Dzhilavyan et al. Yad. Fiz. **51**, 336
[90 Kä]. F. Käppeler et al., *s-Process Nucleosynthesis: Classical Approach and Asymptotic Giant Branch Models for Low-mass Stars,* Astrophys. J. **354**, 630
[90Kl]. N. Klay, F. Käppeler, H. Beer, G. Schatz, *^{176}Lu – An s-Process Thermometer,* Proc. 7th Int. Symp. Capture γ-Ray Spectroscopy, Asilomar, California, 1
[90La]. L. Lakosi, Zs. Németh, I. Pavlicsek, Á. Veres, *Novel Non-destructive Methods for Verification of Spent Fuel Using Neutron and Gamma Activation of Indium,* Kernenergie, **33**, 77
[90Le]. K. L. Lesko, *Changes in Nuclear Decay Rates as a Result of High Stellar Temperature: ^{148}Pm and ^{176}Lu,* Proc. AIP Conf. No. **238**, 861
[90Lo]. M. A. Lone, *Induced Radioactivity from Industrial Radiation Processing,* Nucl. Instr. Meth. **A299**, 656
[90Po]. V. Ponomarev, A. P. Dubenskiy, V. P. Dubenskiy E. A. Boykova, *The Anomaly in the Energy Dependence of the Yield of the Isomer in ^{81}Br by Photoexcitation as Manifestation of Nuclear Intermediate Structure,* J. Phys. G: Nucl. Phys. **16**, 1727
[90Sá]. J. Sáfár, L. Lakosi, *Quantitative Nondestructive Assay of Uranium Bearing Fuel Rods by High Resolution Gamma Spectrometry,* Kernenergie, **33**, 82
[90Ta]. N. C. Tam, A. Veres, I. Pavlicsek, L. Lakosi, *Excitation of 8.85 d ^{129m}Xe and 11.9 d ^{131m}Xe by ^{60}Co Irradiation,* J. Phys. G. Nucl Phys.**16**, 1215
[91Ca1]. J. J. Carroll et al., *Excitation of $^{123}Te^{m}$ and $^{125}Te^{m}$ through (γ, γ') Reactions,* Phys. Rev. **C43**, 897
[91Ca2]. J. J. Carroll et al., *Photoexcitation of Nuclear Isomers by (γ, γ') Reactions,* Phys. Rev. **C43**, 1238
[91Kl]. N. Klay, F. Käppeler, H. Beer, G. Schatz, *Nuclear Structure of ^{176}Lu and Its Astrophysical Consequences I, Level Scheme of ^{176}Lu and II, ^{176}Lu, a Thermometer,* Phys. Rev. **C44**, 2801; **C44**, 2839
[91La1]. L. Lakosi, J. Sáfár, Á. Veres, T. Sekine, K. Yoshihara, *Gamma-Induced ^{115m}In Production in the 0-50 MeV Energy Range,* Proc. Nucl. Data, ed. M. Qaim, Jülich, Springer-Verlag, Berlin, 672
[91La2]. L. Lakosi, I. Pavlicsek, Á. Veres, *Photoexcitation of 3.7 h $^{176}Lu^{m}$ and Its Astrophysical Consequences,* Acta Phys. Hung. **69,** 169
[91Le]. T. Lesko, E. B. Norman, R. M. Larimer, B. Sur, C. B. Beausang, *^{176}Lu: An Unreliable s-Process Chronometer,* Phys. Rev. **C44,** 2850
[91Ne]. P. von Neumann-Cossel et al., *Comment on "Resonant and Nonresonant Contribution to the Photoactivation of ^{111}Cd",* Phys. Rev. **C44**, 554
[91Sá]. J. Sáfár, H. Kaji, K. Yoshihara, L. Lakosi, Á. Veres, *$^{103}Rh^{m}$ Production by Inelastic Gamma Scattering in the Giant Dipole Resonance Region,* Phys. Rev. **C44**, 1086
[91Se]. T. Sekine, K. Yoshihara, L. Lakosi, Zs. Németh, Á. Veres, *Integral Cross Section of the $^{99}Tc(\gamma, \gamma')^{99}Tc^{m}$ Reaction in the 15–50 MeV Energy Region,* Appl. Radiat. Isot.**42**, 141

[91Ta]. N. C. Tam, Á. Veres, I. Pavlicsek, L. Lakosi, *Excitation* of ^{83m}Kr by ^{60}Co Irradiation, Appl. Radiat. Isot. **42**, 431

[91Zh]. W. R. Zhao, F. Käppeler, *Stellar Production Cross Section of $^{176}Lu^{m}$*, Phys. Rev. **C44**, 506

[92Ca]. J. J. Carroll et al., *Reply to "Comment on Resonant Excitation of the Reaction $^{180}Ta^{m}(\gamma, \gamma')^{180}Ta$"*, Phys. Rev. **C45**, 470

[92Co]. C. B. Collins, et. al., *Common Threshold and the Role of Deformations in the Photoexcitation of Isomers,* Phys. Rev. **C46**, 952

[92Ke]. S. E. Kellogg, E. B. Norman, *Isomer-to-Isomer Beta Decay of $^{180}Hf^{m}$ and the Nucleosynthesis of $^{180}Ta^{m}$*, Phys. Rev. **C46**, 1115

[92La]. L. Lakosi, Á. Veres, N. C. Tam, I. Pavlicsek, *Determination of the Production Rate of Low Intensity Isomeric Transitions*, Nucl. Instr. Meth. PRS. **A312**, 17

[92Me]. C. A. Meegan et al., *Spatial Distribution of γ-Ray Burst Observed by BATSE,* Nature, **355**, 143

[92Neu]. P. von Neumann-Cosel et al., *Resonant Photoexcitation of $^{115}In^{m}$ as a Test Case,* Phys. Lett. **B266**, 9

[92Né1]. Zs. Németh, *"Comment on Resonant Excitation of the Reaction $^{180}Ta^{m}(\gamma, \gamma')^{180}Ta$",* Phys. Rev. **C45**, 467

[92Né2]. Zs. Németh, T. Belgya, S. W. Yates, Ch. Theiss, F. Käppeler, *Nucleosynthesis in the Cd-In-Sn Region,* Proc. Symp. On Nucl. Astrophys. Karlsruhe, Germany, 1992, 1993 IOP Publishing Ltd, 219

[92Né3]. Zs. Németh, F. Käppeler, G. Reffo, *Origin of ^{180m}Ta and the Temperature of the s-Process,* Astrophys. J. **392**, 277

[92Sc]. M. O. Scully, *From Lasers and Masers to Phaseonium and Phases,* Phys. Rep. **219**, 191

[93Bi]. I. Bikit et al., *Depopulation of the 6 h Isomeric State of the ^{99}Tc by Photon Irradiation,* J. Phys. G. **19**, 1359

[93Ca1]. J. J. Carroll et al., *Intermediate Structure in the Photoexcitation of $^{77}Se^{m}$, $^{79}Br^{m}$ and $^{137}Ba^{m}$,* Phys. Rev. **C48**, 2238

[93Ca2]. A. G. W. Cameron, in Protostars and Planets, eds. Levi & Lunine, Univ. Ariz.

[93Hu]. M. Huber et al., *Structure of Intermediate States in the Photoactivation of the ^{89}Y Isomer,* Nucl. Phys. **A559**, 253

[93Ju]. M. Jung et al., *First Observation of Bound-State β^{-} Decay,* Nucl. Phys. **A553**, 309

[93La1]. L. Lakosi, J. Sáfár, *(γ, γ') Reaction Studies by Excitation of Isomers Around and Above Neutron Binding Energy,* Proc. Capt.γ-Ray Spectr., ed. Jean Kern, Friburg, World Sci. Singapore, 611

[93La2]. L. Lakosi, J. Sáfár, Á. Veres, T. Sekine, K. Yoshihara, *Photonuclear Reaction on ^{99}Tc Isomer Excitation and De-excitation, Implication in Nucleosynthesis,* Radiochem. Acta **63**, 23

[93La3]. L. Lakosi, J. Sáfár, Á. Veres, T. Sekine, H. Kaji, K. Yoshihara, *Photoexcitation by Gamma Ray Scattering Near Threshold and Giant Dipole Resonances,* J. Phys. G. Nucl. Phys. **19**, 1037

[93Ma1]. V. M. Mazur, I. V. Sokolyuk, Z. M. Bigan, I. Y. Kobal, *Nuclear Isomer Excitation Cross Section in $(\gamma, \gamma')^{m}$ Reactions at 4–15 MeV,* J. Nucl . Phys. **56/1**, 20

[93Ma2]. V. E. Makarenko, G. B. Yankov *(Isomeric States of Nuclei) (in Russian),* J. Nuclear Physics **56/12,** 1

[93Sá1]. J. Sáfár, L. Lakosi, *Photoexcitation of ^{93m}Nb in the 5- to 30-MeV Gamma Energy Range,* Nucl. Sci. Engin. **115**, 297

[93Sá2]. J. Sáfár, L. Lakosi, Á. Veres, T. Sekine, K. Yoshihara, *Photoexcitation of Metastable States in ^{111}Cd, ^{113}In and ^{115}In in the 5–30 MeV Energy Range,* Proc. 8th Int. Simp. Capt. γ-Ray Spectr. ed. Jean Kern, Friburg, World Sci., Singapore, 629

[93Sc]. V. Schönfelder et al., *Instrument Description and Performance of the Imaging Gamma-Ray Telescope COMPTEL Aboard the Compton Gamma-Ray Observatory,* Astrophys. J. Suppl., **86**, 657

[94Fi]. C. E. Fichtel et al., *The First Energetic Gamma-Ray Experiment Telescope (EGRET) Source Catalogue,* Astrophys. J. Suppl., **94**, 551

[94Ge]. W. Geiger et al., *Low-lying Dipole Excitations in the Isotopes $^{113,114}Cd$.* Nucl. Phys. **A580**, 263

[94Kn]. D. A. Kniffen, E. Chipman, N. Gehrels, *Frontier of Space and Ground-Based Astronomy,* eds W. Wamsteker et al., Kluwer Academic Publ., Netherlands, 5

[94La]. L. Lakosi, I. Pavlicsek, N. C. Tam, *Auto-coincidence Counting of Low Intensity Isomeric Transition,* Nucl. Instrum. Meth. Phys. Res. **A339,** 226

[94Né1]. Zs. Németh et al., *Complete Spectroscopy of ^{113}Cd,* Proc. 8th Int Symp. Capt. γ-Ray Spectroscopy, ed. J. Kern, World Sci. Singapore, 314

[94Né2]. Zs. Németh, F. Käppeler, Ch. Theis, T. Belgya, S. W. Yates, *Nucleosynthesis in the Cd-In-Sn Region,* J. Ap. **426**, 357

[94Neu]. P. von Neumann-Cosel et al., Nucl. Instrum. Meth. Phys. Res. **A388**, 425

[94Sc]. C. Schlegel et al., *Depopulation of $^{180}Ta^{m}$ by Coulomb Excitation and Possible Astrophysical Consequences,* Phys Rev. **C50**, 2198

[94Se]. T. Sekine, K. Yoshihara, J. Sáfár, L. Lakosi, A. Veres, *^{95}Tc and ^{96}Tc Production by (γ, xn) Reactions,* J. Radioanal. Nucl. Chem. Lett. **186**, 165

[95Co]. C. B. Collins, J. J. Carroll, Laser Physics (Moscow) **5**, 209

[95Cou]. R. G. Coussement et al., Laser Physics (Moscow) **5**, 292

[95Kh]. N. X. Khan, L. Lakosi, I. Pavlicsek, *Excitation of $^{77}Se^{m}$, $^{79}Br^{m}$ and $^{191}Ir^{m}$ by ^{137}Cs Gamma Rays,* Phys. Rev., **C51**, 1676

[95Ko]. O. Kocharovskaya, Laser Physics, **5**, 284

[95La1]. L. Lakosi, N. X. Khan, N. C. Tam, J. Sáfár, I. Pavlicsek, *Photoexcitation of Isomers by Bremsstrahlung of 4 MeV Electrons,* Appl. Radiat. Isot. **46**, 435

[95La2]. L. Lakosi, N. X. Khan, N. C. Tam, I. Pavlicsek, A. Petõ, *Isomer Excitation by High Activity ^{137}Cs and ^{60}Co Sources and Low Energy X-Rays,* Appl. Radiat. Isot., **46**, 433

[95La3]. L. Lakosi, N. C. Tam, I. Pavlicsek, *Photoexcitation of $^{189}Os^{m}$ and $^{193}Ir^{m}$ I.,* Phys. Rev. **C52**, 1510

[95La4]. L. Lakosi, N. C. Tam, I. Pavlicsek, Á. Petõ, *Photoexcitation of $^{189}Os^{m}$ and $^{193}Ir^{m}$ II.,* Phys. Rev. **C52**, 1516

[96Fi]. R. B. Firestone et al., *8th Table of Isotopes,* J. Wiley and Sons, Inc., New York, 1996

[96Ga]. Yu. P. Gangrsky, A. P. Tonchev, N. P. Balabanov, *Excitation of Isomeric States in the Photonuclear Reactions,* Fiz. Elementar. Chashtic Atom Jadr. **27**, 1043

[96Ki]. R. L. Kinzer et al., *OSSE Observations of Positron Annihilation in the Galactic Plane,* A&A Suppl. **120**, 317

[96La]. L. Lakosi, *Production of Long-lived Isomers,* Z. Phys. **A356**, 155

[97Ar]. T. Arisawa et al., *Separation of an Isotope as a Precursor of a Gamma-Ray Laser Medium*, Hyp. Int. **107**, 116

[97Be]. A. G. Below, Yu. P. Gangrsky, A. P. Tonchev, P. Zuzan, *Excitation of the High-spin ^{180}Hf Isomer and De-excitation of the ^{180}Ta in (γ,γ′) Reactions,* Hyp. Int. **107**, 167

[97Bi]. I. Bikit et al., *New Gamma Transition in the Decay of ^{137}Cs,* Proc. 9th Int. Symp. Capture Gamma-ray Spectroscopy, Eds. G. L. Molnár, T. Belgya, Zs. Révay, Springer, Budapest, 303

[97Ca]. J. J. Carroll, C. B. Collins, *Limits on Spurious Contributions to Integrated Cross Sections for Photoexcitation and De-excitation of Isomers,* Hyp. Int. **107**, 149

[97Co1]. C. B. Collins, J. J. Carroll, *Progress in the Pumping of a Gamma Laser,* Hyp. Int. **107**, 3

[97Co2]. C. B. Collins, J. J. Carroll, Yu. Ts. Oganessian, S. A. Karamian, *Evidence for K Mixing in ^{178}Hf,* Hyp. Int. **107**, 141

[97Ho]. G. R. Hoy, *Time-Domain, Nuclear-Resonant, Forward Scattering: the Classical Approach,* Hyp. Int. **107**, 381

[97Ka]. S. V. Karjagin, *Gamma-Ray Solid Laser: The Heat Problem and Means of Solution; Realisation of Pumping,* Hyp. Int. **107**, 449; 465

[97Ko] O. Kocharovskaya, *Lasing Without Inversion: Problems and Prospects,* Hyp. Int. **107**,187

[97La]. L Lakosi, J. Sáfár, H. Kaji, *Excitation of ^{180m}Hf by 15–30 MeV Bremsstrahlung,* Proc. 9th Int. Symp. Capture Gamma-ray Spectroscopy, eds G. L. Molnár, T. Belgya, Zs. Révay, Springer, Budapest, 316

[97Neu]. P. von Neumann-Cosel et al., *Coulomb Excitation of $^{180}Ta^m$ and Its Consequences for a Possible s-Process Production,* Nucl. Phys. **A621**, 278c

[97Og]. Yu. Ts. Oganessian, et al., *K-mixing in Nuclear Reactions,* Hyp Int. **107**, 43; *Production, Chemical and Isotopic Separation of the $^{178m2}Hf$ High-spin Isomer,* Hyp Int. **107**, 129

[97Sc]. V. Schönfelder, *Astronomical Gamma-Ray Line Spectroscopy with the Compton Observatory,* Proc. 9th Int. Symp. Capture Gamma-ray Spectroscopy, eds G. L. Molnár, T. Belgya, Zs. Révay, Springer, Budapest, 582

[97Sh]. R. N. Shakamuratov, *Lasing Without Inversion Due to Cooling System,* Hyp. Int. **107**, 205

[97Ve]. Á. Veres, *On the Half-life of Activation Levels of ^{111}Cd and $^{113,115}In$ in the 1.38–2.75 MeV Energy Range,* Proc. 9th Int. Symp. Capture Gamma-ray Spectroscopy, eds G. L. Molnár, T. Belgya, Zs. Révay, Springer, Budapest, 284

[97Wi]. M. Wiescher, H. Oberhummer, *Nuclei in the Cosmos: Present Status and the Future of Nuclear Astrophysics,* Nucl. Phys. News, **Vol. 7, No. 3**, 11

[98Al]. R. A. Alpher, R. Herman, *On the Density Ratio Omega in the Standard Big Bang Model,* APH N.S. Heavy Ion Phys. **8**, 185

[98So]. J. C. Solem, *Status of Research in Gamma-Ray Lasers,* APH N.S. Heavy Ion Phys. **7**, 185

[99Be]. D. Belic et al., *Photoactivation of $^{180}Ta^m$ and Its Implications for the Nucleosynthesis of Nature's Rarest Naturally Occurring Isotope,* Phys. Rev. Lett. **83**, 5242

[99Bi1]. I. Bikit, L. Lakosi, J. Sáfár, Lj. Conkic, *Depopulation of $^{180}Ta^m$ by Bremsstrahlung,* Phys. Rev. **C59**, 2272

[99Bi2]. I. Bikit, L. Lakosi, J. Sáfár, Lj. Conkic, *De-excitation of ^{180m}Ta by ^{60}Co Gamma Rays,* Astrophys. J. **522,** 419

[99Co]. C. B. Collins et al., *Accelerated Emission of Gamma Rays from the 31-yr Isomer of ^{178}Hf Induced by X-Ray Irradiation*, Phys. Rev. Lett. **82,** 695

[99Ha]. R. C. Hartman et al., *The Third EGRET Catalogue of High-Energy Gamma-Ray Sources,* Astrophys. J. Suppl., **123**, 79